T0205732

Lecture Notes in Chemistry

Volume 101

The Lecture Notes in Chemistry

The series Lecture Notes in Chemistry (LNC) reports new developments in chemistry and molecular science-quickly and informally, but with a high quality and the explicit aim to summarize and communicate current knowledge for teaching and training purposes. Books published in this series are conceived as bridging material between advanced graduate textbooks and the forefront of research. They will serve the following purposes:

- provide an accessible introduction to the field to postgraduate students and nonspecialist researchers from related areas,
- provide a source of advanced teaching material for specialized seminars, courses and schools, and
- be readily accessible in print and online.

The series covers all established fields of chemistry such as analytical chemistry, organic chemistry, inorganic chemistry, physical chemistry including electrochemistry, theoretical and computational chemistry, industrial chemistry, and catalysis. It is also a particularly suitable forum for volumes addressing the interfaces of chemistry with other disciplines, such as biology, medicine, physics, engineering, materials science including polymer and nanoscience, or earth and environmental science.

Both authored and edited volumes will be considered for publication. Edited volumes should however consist of a very limited number of contributions only. Proceedings will not be considered for LNC.

The year 2010 marks the relaunch of LNC.

More information about this series at http://www.springer.com/series/632

Ewa Bulska

Metrology in Chemistry

Ewa Bulska
University of Warsaw
Warsaw, Poland

ISSN 0342-4901 ISSN 2192-6603 (electronic)
Lecture Notes in Chemistry
ISBN 978-3-030-07576-7 ISBN 978-3-319-99206-8 (eBook)
https://doi.org/10.1007/978-3-319-99206-8

English translation of the 2nd original Polish edition published by Wydawnictwo Malamut, Warszawa, 2012

This Springer imprint is published by the registered company Springer Nature Switzerland AG
The registered company address is: Gewerbestrasse 11, 6330 Cham, Switzerland

Preface

My primary motivation for putting some crucial issues related to chemical measurements in the form of a comprehensive textbook was my involvement in training courses related to those topics, in addition to the fact that I was actively involved in accreditation activities around Europe. The original idea for this book came about during my active involvement in the TrainMiC® project, which is related to knowledge dissemination on good practice in obtaining valid results in chemical measurements, financed by EU funds.

Having been involved in analytical chemistry research as well as the development of novel methodology and procedures, I am convinced that the issue of applying all relevant metrology principles needs more awareness from those involved in the field of analytical chemistry. Having been active in the curricula of various training courses for students and professionals, in support of accreditation processes, and in national as well as international organizations focused their activities on the quality of measurement results and its legal validity, I am convinced that knowledge in metrology—apart from its use in routine laboratories—is a fascinating field that is developing with a great deal of future potential. This is why I hope that *Metrology in Chemistry* will help to impart the joy of applying good metrological practice and challenge its modern understanding among the readers of this book.

Metrology in chemistry is recognized as a discipline that combines knowledge on analytical chemistry, physical chemistry, statistics and general metrology, as well as legal regulation, on a worldwide level. Although in-depth understanding, as well as practical use of this knowledge, have developed dynamically over recent years, there is still need to spread awareness on good practice in chemical measurement in academia as well as in routine testing and calibration laboratories. The fundamental principles of metrology do not vary between disciplines; however, the specificity of the measurements matters in terms of how they are implemented in practice. Thus, this book aimed to comprehensively cover the up-to-date knowledge on how to carry out the measurements of chemical properties of various objects, referring, when the opportunity arises, to other related disciplines. I wish for this book to be of use to those who perform chemical measurements and for those who use the results of chemical measurements to make decisions or establish regulations.

The first edition (in Polish) was issued in 2008 by MALAMUT Publishing House (www.malamut.pl), then four years later, when the first edition had sold out, the second, revised edition was published in 2012. In the ten years since the publication of the first edition of this monograph, the understanding and application of metrology in chemistry has undergone remarkable development. It was necessary to revise numerous chapters substantially; this was also due to the new edition of ISO 17025, which appeared at the end of 2017. This book is the English-language version of its Polish forerunner *Metrologia chemiczna*, which is now in its third, updated edition.

When preparing the first edition and then the following editions of *Metrology in Chemistry,* I benefited from the support and experience of a significant number of colleagues. I have been fully convinced of the benefits of collaboration and of holding chemical/metrological friendships. I experienced strong support from a large number of friends who were willing to share their knowledge and experience, all benefiting the content of this book. My sincere thanks are directed to those who contributed strongly to the shape of this book. To Prof. Adam Hulanicki from the University of Warsaw, for his critical comments and wise advice. To Dr. Anna Ruszczyńska from the University of Warsaw for her enthusiasm in preparing all illustrations included in this book. I appreciate the support from Dr. Piotr Bieńkowski, Editor of MALAMUT Publishing House for his full support during the time the first and the second Polish edition of *Metrology in Chemistry* were brought out on the market, as well as for pointing me in the right direction for preparing the English translation. Last but not least, I want to express my pleasure that my book on *Metrology in Chemistry* has been accepted for translation into English by Springer.

I am aware that this book would not have appeared in its current form without the support of many of my friends, colleagues and authorities. However, I alone am responsible for any inadequacies and imperfections—after all, this is the first textbook purely dedicated to metrology in chemistry.

Warsaw, Poland Ewa Bulska

Contents

Abbreviations and Acronyms

AOAC	Association of Official Analytical Chemists
BIPM	Bureau International des Poids et Mesures (International Bureau of Weights and Measures)
CCQM	Comité Consultatif pour la Quantité de Matière (International Committee for Weights and Measures)
CEN	European Committee for Standardization
CGPM	Conférence Générale des Poids et Mesures (General Conference on Weights and Measures)
CIPM	Comité international des poids et mesures (International Committee for Weights and Measures)
CITAC	Co-operation on International Traceability in Analytical Chemistry
CMC	Calibration and Measurement Capability
CODATA	Committee on Data for Science and Technology
CRM	Certified Reference Materials
EA	European co-operation for Accreditation
EFTA	European Free Trade Association
ERM	European Reference Material
Eurachem	European Network of Analytical Chemistry Laboratories
EURAMET	European Collaboration in Measurement Standards
IAEA	International Atomic Energy Agency
ICTNS	International Committee on Terminology, Nomenclature and Symbols
IEC	International Electrotechnical Commission
ILAC	International Laboratory Accreditation Cooperation
ILC	Inter-laboratory comparisons
ISO	International Organization for Standardization
IUPAC	International Union of Pure and Applied Chemistry
IUPAP	International Union of Pure and Applied Physics
LoD	Limit of Detection
LoQ	Limit of Quantitation

MRA	Mutual Recognition Arrangement
NIST	National Institute of Standards and Technology
NMIs	National Metrology Institutions
OECD	Organization for Economic Co-operation and Development
OIML	Organisation Internationale de Métrologie Légale (International Organization of Legal Metrology)
PT	Proficiency Testing
QA	Quality Assurance
QC	Quality Control
QMS	Quality Management System
RM	Reference Material
RMO	Regional Metrology Organization
RMP	Reference Material Producer
SI	Système Internationale — d'unités (International System of Units)
SRM	Standard Reference Material
VIM	Vocabulaire International de Métrologie (International Vocabulary of Metrology: Basic and General Concepts and Associated Terms)
WELMEC	European Cooperation in Legal Metrology
WHO	World Health Organization

Introduction: A Brief History of Measurement

The genesis of measurement is part of the natural human drive to compare, for example, the duration of a process, the distance traveled, the temperature or the taste of food. Comparing always encompasses referring to a standard. Claiming that a distance traveled was longer or shorter has a reference to some commonly used distance, in relation to which our pathway took more or less time. Therefore, each community has had to work out a system of standards, which were set up as a commonly accepted reference.

The great meaning of measurements is supported by numerous records in civil and religious documents relating to the functioning of society; for example, in the Code of Hammurabi, in the Bible or in the Koran. In the Bible, there are many warnings against the sin of giving short measure. The first available data of Polish surveyors can be found in the Bull of Gniezno from the year 1136. Surveyors were engaged to determine the borders of settlement fields, using a rope and a pole as measuring devices. Units of measurement were often created on the basis of items used in practice, hence the ancient length units of rope and pole. Historians recall many practical ways of conducting measurements, since human invention in creating convenient units of measurement from everyday life was unlimited (Fig. 1). Measurements were often based on the observation of nature; for example, the day was the time period between the sunrise and sunset. Ancient prayers were often used as a unit of time—for counting time sufficient to, say, the prayer Ave Maria. When a prayer was completed at a common rate (e.g., 20 s), this was agreed upon as a common unit of time. Time was sometimes counted by the period needed for completing the burning of commonly used candle.

For measuring length, units referred to the human body: the foot (the length of a foot), span (related to the human hand—the span between spread fingers), elbow (the length of the forearm) or inch (the thickness of the thumb) were introduced. It is, therefore, obvious that the size of the unit depended on the person whose body part served as a reference. A comparison of length was therefore only possible when both interested parties referred to the foot length of the same person, i.e. a common standard. What about the unit of a stone's throw? Questions immediately arise: who

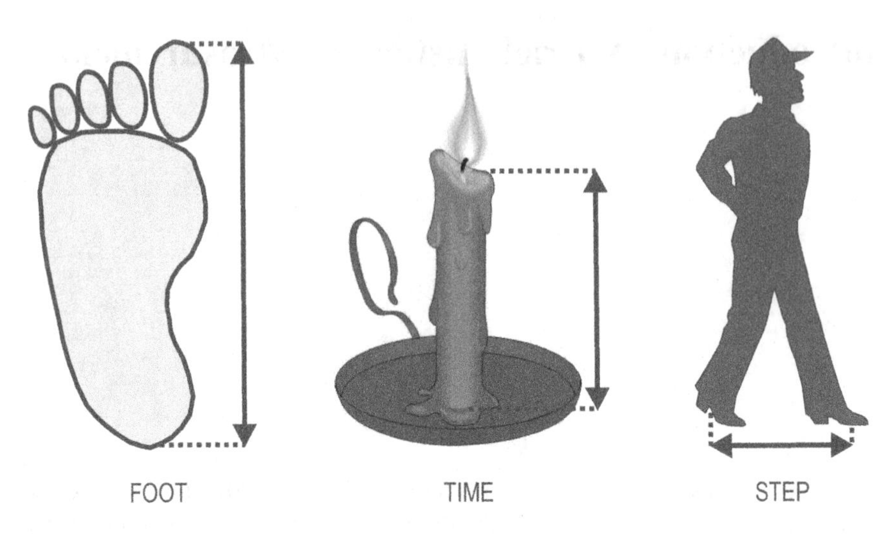

Fig. 1 Examples of sources of primarily used unit of measurements

threw the stone? What stone was it? What was the weather like? Does that not remind us of the need to describe the conditions in which the measurement was carried out—that is, a measuring procedure?

By tracking the formation of units of measurement with time, we can see, that apart from anthropological units or those derived from the observation of nature, humans also began to use technological achievements for that purpose. Mastering the production of glass resulted in the introduction of the unit 'slab,' which corresponded to the size of the commonly used rolling mill table in the glass factory. The commonly used term in technology, 'horsepower,' corresponded to the power of a steam machine that carried out work replacing one live horse.

Primarily used units of measurement were set up locally and accepted within a given community. In those times, the separateness of metrics used in different countries/territory or even in various districts was considered as proof of power and a testament to freedom. However, as the cooperation between communities developed, such separateness became a hindrance. As the free trade, communication and industry grew internationally, the variety of units of measurement became more and more of a problem, and their unification and creating uniform units by interested parties became a pressing issue. As a result of the efforts of the French Academy of Science, units of length and mass were unified and introduced in France: the units of meter and kilogram, respectively. Those actions and the benefits resulting from them were appreciated by other governments, which soon resulted in the adoption of a consistent system of units, named the International System of Units (abbreviated as SI from the French, *Le Système Internationale—d'unités*). Nowadays days, SI is recognized as the modern and most widely used system of measurements.

An advantage of the international system of units, accepted finally in 1960 by the General Conference on Weights and Measures (CGPM; *Conférence Générale des Poids et Mesures*), is that it was accepted by a number of countries and is commonly used in the international forum.[1] It is worthwhile to add that the SI was introduced as a result of a convention between the governments of the signatories' countries—an agreement that was drawn regardless of political divisions and borders. An inestimable value of using a consistent system of units of measurement is the ease of the exchange of wares and services. Nowadays, it is hard to imagine societies functioning without a uniform system of units of measurement. It would not be possible to carry out production, trade or to care for nature or food, without conducting measurements.

[1]As of March 23th, 2018, there are 59 Member States and 42 Associates of the General Conference.

Chapter 1
Introduction to Metrology

"Count what is countable, measure what is measurable and what is not measurable, make measurable." *Galileo Galilei (1564–1642)*

1.1 Metrology in General

The name 'metrology' derived from the fusion of two Greek source words: *métron* meaning measurement; and *lógos* meaning science.

As defined by the International Bureau of Weights and Measures (BIPM; *Bureau International des Poids et Mesures*): "Metrology is the science of measurement, embracing both experimental and theoretical determinations at any level of uncertainty in any field of science and technology." It establishes a common understanding of units, which become crucial to all human activity. The source word *métron* can be found, among others, in the term 'metronome,' which is a device used in music for the precise and loud measuring of musical tempo, with the help of the movement of a pendulum. The word 'metronome' is also used in the theory of science information as the smallest unit of metrical information obtained after an experiment is conducted.

The need to establish measuring methods and tools can be noticed from the earliest period of the development of civilization. To be able to survive in a hostile environment, humans had to learn to estimate (to measure) the distance, speed, mass, time, strength, temperature, and so on. In the beginning, the assessment of the properties of the surrounding nature was conducted through senses and in relation to locally agreed units. As it was emphasized in the introduction, a huge disadvantage of those early systems, was the subjectivity of the measurements, which stemmed from the fact that the result of the measurement depended on the specifics of the measuring unit used. As an example, one can recall the 'foot,' used as a measuring unit, which illustrates how much the result of measurement was dependent on the 'owner' of the foot used to compare the length of other objects (Fig. 1.1).

E. Bulska, *Metrology in Chemistry*, Lecture Notes in Chemistry 101,
https://doi.org/10.1007/978-3-319-99206-8_1

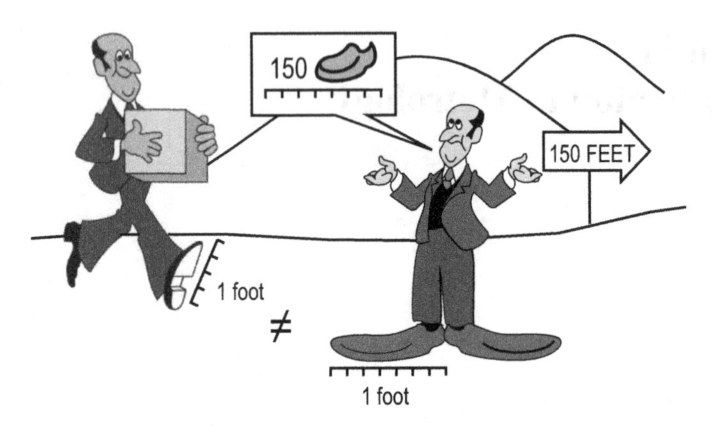

Fig. 1.1 The natural variation of the 'definition' of foot as a unit of length

The ability to use units of measurement and the development of methods and tools used to make said measurements in everyday life is called *surveying*—the use of metrology in practice. In common understanding, '*surveyor'* is the name of a specialist who uses measurement tools in practice. A surveyor measures, for example, an area of land and then prepares maps of the measured areas. Whereas *metrology* deals with the theory of measurement and at its core is the theoretical description of the measurement principles. It is worth noticing that the term 'measuring' is used in many contexts; for example, measuring tools, measuring methods, measuring tape, measuring team. Nowadays, we also consider also metrologists—that is, specialists who take measurements. In publications, we may also encounter terms relating to specific areas of measurement; for example. the temperature measurement is called 'thermometry' and those who take such measurements are referred to as 'thermometrists.'

It is commonly assumed that measurement is about determining the numerical value that quantifies the property or properties of the studied object. The valuation of a given property requires comparing it to the standard of the measured value, which allows it to be assigned with a numerical value expressed in the unit of that standard. Hence the establishment of a uniform system of units of measurement that was and still is crucial for the development of trade and economy. Apart from this, technological advances have enabled many new measuring methods and tools to develop, and thus the values that not so long ago were impossible to measure became measurable. An example from recent years is the development of *nanometrology*, the area of science that makes measurements on the nano-scale, thanks to which human have gained new potential areas for study. It should be noted however, that the significant development of measuring tools was connected with subsequent periods of the development of civilization. Moreover, increasing awareness that each measurement is subjected to a wide variety of possible errors has pushed development of metrology towards the ability to evaluate such errors.

Measurements have been carried out and used by humans for thousands of years. In order to make them relevant, a property values are compared with values of those properties for accepted standards, accompanied by a priori established limits, within which the properties can differ as a result of errors in measurement. In practice, measuring includes establishing accepted standards, comparing the properties of standards to the properties of studied objects, and establishing the accepted range of variation of results of individual measurements. Without measurements, it is hard to imagine the development of industry, transport, trade, as well as scientific fields such as physics, chemistry, mechanics, astronomy, thermodynamics, electronics, and many others, in which measurement allows verification of scientific hypotheses.

The development of technology often requires an extensive volume of information. As such, it is not just a single measurement that is required but also information on a numerous set of properties of an object or of a large number of objects. As a result, there is also a need to determine and register not only the properties of the object, but also the its variability with time, the direction of the changes, predictions of the properties, the recognition of the images, and so on. The results of measurements of various properties were always the basis of settlement between people and were a part of everyday practice in society. Hence for a long time, people did not feel the need to create the theoretical superstructure. In practice, all they needed was the empirical knowledge of how to perform measurements to ascertain the numerical characterization of given items.

In fact, metrology, the science of measurements, has a crucial role in the development of society. Like any science, it is not only concerned with material objects but also with abstractions relating to the entire class of object properties. Metrology as such is a relatively newly recognized science, and at its core is the theoretical—mostly mathematical—establishment of measuring principles. Metrology, dealing with the theory of measurements, aims to discover measuring challenges and new cognitive issues. Defining units of measurement, implementation of units of measurement, and last but not least setting chains between standards are among the most important tasks of metrology.

Surveying, also known as a practical activity towards measuring the properties deals with the practical aspects of metrology—the technicalities of the evaluation of the material world's overview. It is, hence, the part of technology of a service character, which determines the development of science and technology to a great extent, and it involves almost all aspects of human activity.

- **Measurement is a technique of evaluating the properties of objects from the material world.**
- **Metrology is the science around the principles of measurement.**

In this book, most topics refer exclusively to metrology, with a strong focus on the measurement of chemical properties of objects. In the following chapters, the scope of discussion around metrology includes the theory of the units of measurement and the methods of their reproduction.

In practice, the following various branches of metrology are distinguished:

- General metrology: dealing with all items related to measurements that are common to all fields; for example, the theory of the units of measurement and the evaluation of the general properties of measuring devices.
- Scientific metrology: dealing with the definition of the units of measurement and the development of standards and measuring tools at the highest metrological level.
- Theoretical metrology: dealing with theoretical questions of measurements (e.g., measuring errors) and measuring techniques.
- Industrial metrology: dealing with ensuring the proper functioning of measuring devices used in the industry.
- Legal metrology*: dealing with the issues of units of measurement, measuring methods and tools from the perspective of the officially defined technical and legal requirements.
- Military metrology: dealing with ensuring accuracy and reliability of measurements in all areas connected to the national defence.

* Legal metrology is the application of legal requirements to measurements and measuring instruments. Legislation on measurements and measuring instruments is required in many cases, as well as when there is a need to protect both the buyer and the seller in a commercial transaction, or where measurements are used to apply a sanction. Virtually all countries provide such protection by including metrology in their legislation—hence the term 'legal metrology.' International Organization of Legal Metrology (OIML; *Organisation Internationale de Métrologie Légale*).

The commonly used term 'applied metrology' refers to a specific kind of property measurement or to measurements performed in specific areas. Applied metrology is divided into areas related to the specific kind of measured values (e.g., electric metrology, metrology of length, metrology of time) and to its various uses (e.g., workshop metrology, technical metrology, medical metrology, energetic metrology or the abovementioned military metrology).

Apart from that, metrology is divided depending on the qualities that are measured. In the early period of development of general metrology, the measuring principles introduced applied only to physical properties, and with time those principles were also applied to chemical properties. Those specific rules regarding conducting measurements of chemical properties were referred to as '*chemical metrology*.' The metrology principle regarding measuring chemical properties obviously does not differ from those valid in measuring physical properties. The defining of the measuring area stems largely from the specific nature of measurements in the given field and from the fact that practical application of principles differers. Along with the development of other fields, the principles of measurements enter the area of microbiological metrology, biological metrology, metrology of nanomaterials (nano-metrology), and so on. Wherever measurements are carried out, the implementation of metrology principles should be seen as legitimate, and so should the adoption of metrology principles in practice in the given area.

1.2 Specificity of Chemical Metrology

Chemical measurements performed in many fields are used for many crucial decisions regarding, for example, safety and quality of life. The results of clinical measurements are used by medical doctors to make decisions regarding the status of our health and the possible need for medical treatment is worked out on the basis of those results. Information on the composition of foodstuffs are important in respect to being allowed in the marketplace, and ensuring food safety for consumers. In addition, those are not the only areas in which chemical measurements are used; also important are the anti-doping tests in sport, the quality control tests in production processes, monitoring of the environment, and many others. Due to the need to carry out various chemical measurements, namely as a consequence of the introduction of related legal regulations (within European Union the recommendations stem from, among other things, the requirements described in various directives, e.g., the Water Framework Directive), the number of testing labs that carry out such measurements and the number of people employed has been increasing exponentially in recent years. Hence there is a need to establish and implement common principles of measurements conducted in testing laboratories, in order to ensure the credibility of the results of measurements. The international standard ISO/IEC 17025, 'General Requirements for the Competence of Testing and Calibration Laboratories,' requiring validation of the measuring procedure, highlighted the uncertainty of the result obtained with the use of the procedure and the ensuring of measurement traceability were highlighted. Those three parameters: validation, uncertainty and traceability are considered to be a pre-requirement for obtaining sound results, hence their relationship with metrology and its principles are closely linked.

What exactly is chemical metrology and what is its position in metrology? It is the metrology principle used for the evaluation of chemical properties of given objects by chemists in the analytical laboratory so as to perform measurements, aiming to determine the qualitative and quantitative composition of samples.

1.2.1 *Calibration of a Measuring Tool*

Before discussing the specifics of chemical measurement, it is worth referring to the term used in the context of measurement traceability:, i.e. 'calibration.'

Calibration: the operation that, under specific conditions, first establishes a relationship between the quantity values with measurements uncertainties provided by measurement standards and corresponding indications with associated measurements uncertainties and, second, uses this information to establish a relationship for obtaining a measurement result from an indication. [Clause 2.39; ISO/IEC Guide 99 (VIM)]

When the name ISO/IEC Guide 99 is used, this refers to the International Vocabulary of Metrology—Basic and General Concepts and Associated Terms (VIM; Vocabulaire International de Métrologie*)*

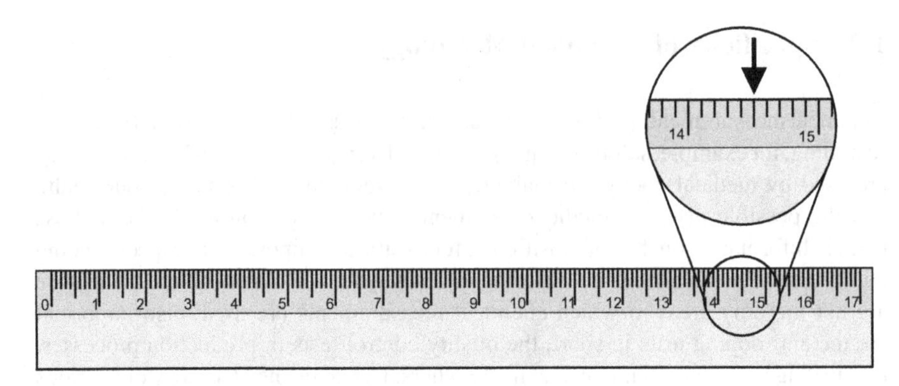

Fig. 1.2 Commonly used ruler for the measurement of length

Originally, the principles of metrology were introduced for the measurement of physical properties (e.g., mass, length, temperature), for which the result depends greatly on the quality of the measuring tool (a tool for measuring mass, length, thermometer), and in most cases does not depend on the kind of object that is measured.

A commonly used tool for length measurement is a linear gauge, popularly known as the ruler. Most commonly, the linear gauge is calibrated in centimetres, and the smallest graduations scale is 1 mm. Using such a linear gauge, the smallest length that can be measured, without using approximation, is a millimetre.

Figure 1.2 shows an enhanced scale of a part of a school ruler. Longer, numbered lines are scaled in centimetres, each is divided into 1-mm graduations.

Note: the precision of the results obtained with a given measuring tools is usually one significant number more than the smallest graduation of its scale.

Using a linear gauge, graduated in intervals of 0.1 cm, the value of the measured property (length) can be read accurately to 0.05 cm. By taping such a ruler to the object, it can be stated that the measured length lies between, for example, 24.4 and 24.5 cm. Then it can be approximated that the measured length lies, for example, in the middle of the smallest (1 mm) graduation, which corresponds to the length of 24.45 cm. If the length is required to be expressed in the basic units of the SI, it should be expressed as 0.2445 m.

As with measuring temperature, the graduation on the glass tube of the thermometer matters. One should be aware that the values assigned to the graduations (the drawn lines) result from calibration, and that the reference thermometer used in the calibration of the working thermometer was part of a traceability chain, which ensures consistency within the reference standard of the SI temperature unit. In chemical measurements, apart from the calibration of measuring instruments as such, the kind of samples analyzed and the method of its preparation prior to conducting measurements are extremely important. With regard to the straightforward examples, when measuring temperature of any liquid, what is important is the calibration of the used thermometer—the kind of liquid being measured does not have a significant influence on the result. In the chemical measurement, when determining the content of a given substance (analyte) in the sample, apart from the need to use the calibrated instrument properly, it is also necessary to establish the influence of the accompanying sample components (the chemical matrix) on the final result. In chemical analyses, the validation of the whole analytical procedure is crucial, and should include sampling, sample preparation, measurement of the given properties, and evaluation of the influence of the matrix on the measurement's result and the evaluation of the influence of all of these parameters on the final test result.

A great deal of discussions and controversies are evoked by the use of the terms 'measurement' and 'test.' They are not synonymous; their meanings differ and their concepts are described in the following documents, respectively: ISO/IEC Guide 99 (VIM) and EA-04/16. The distinction in the meaning of those terms and their correct use is important.

Measurement: the process of experimentally obtaining one or more quantity values that can reasonably be attributed to a quantity. [Clause 2.1; ISO/IEC Guide 99 (VIM)]

Test: technical operation that consists of the determination of one or more characteristics of a given product, process or service according to a specified procedure. [Chapter 5.2; EA—4/16 G:2003 EA guidelines on the expression of uncertainty in quantitative testing]

Following the more detailed description given in EA-4/16 G:2003, it is clear that in general a measurement process yields a result that is independent of the measurement method, while the only difference relates to uncertainty associated with a particular method. In the example of the temperature measurements of a given object, apart from the type of thermometer, the measured values should be the same, varying only with the uncertainty that depends on the performance of the thermometer. By contrast, a test result depends on the method of measurement and on the specific procedure used to determine the given property. As a consequence, different test methods may yield different results. Thus, in the case of measurement procedures, environmental and operational conditions will either be maintained at standardized values or be measured in order to apply correction factors and to express the result in terms of standardized conditions. From the given description it follows that measurement is an integral part of a test.

Accordingly, the responsibility of a chemical testing laboratory is not limited to performing a measurement of the agreed chemical properties of the object delivered by the client but also includes the entire analytical procedure established to be fit for the purpose of the client's need. This requires the designing of a proper technical procedure that is adequate for the objective, for which the measurement results will be used.

It is worth emphasizing here that this book is addressed to those who mostly deal with the measurement of chemical properties. Hence in the following chapters, particular attention will be paid to the metrological aspects of the analytical procedure executed by the chemical laboratory (excluding sampling from the entire object), namely: laboratory sample preparation, measurement, and estimation of the influence of matrix components and the conditions of sample preparation on the measurement result. Of course, aspects related to the calibration of measuring instruments will also be considered. Nevertheless, a lot of attention will be focused on the fact that the result of chemical measurements depends to a great extent on the kind and composition of the investigated objects, and on the method of its preparation before conducting the instrumental measurements.

Example The procedure used for samples preparation: when determining DDT (dichlorodiphenyltrichloroethane) in meat, the use of a calibrated GC-MS (Gas Chromatography Mass Spectrometry) device does not a priori ensure reliable results, since the results depend on, among other things, the procedure and reagents used for the extraction of DDT from the meat sample. Moreover, the conditions in which the process was carried out (e.g., time, temperature) influence the effectiveness of the DDT extraction.

Appropriate definition of the measurand: this belongs to the more important aspects of metrology and is especially important in chemical measurements. The question about the determination of cadmium in the soil can concern both the determination of the total content of the element in the soil sample delivered to the laboratory or the fraction that can be extracted in given conditions. For that reason, it is crucial to identify the measurand and the purpose of conducting the measurements in a given object, so that the comparison of the measurement results refers to the same, clearly defined property.

Measurand: Quantity Intended to Be Measured

NOTE 1: The specification of a measurand requires knowledge of the **kind of quantity**, description of the state of the phenomenon, body, or substance carrying the quantity, including any relevant component, and the chemical entities involved. [Clause 2.3; ISO/IEC Guide 99]

To be deleted Clause 2.3; International Vocabulary of Metrology—Basic and General Concepts and Associated Terms (VIM; Vocabulaire International de Métrologie*)*

Measuring procedures used in the chemical laboratory are usually complex processes (Fig. 1.3), beginning with the sample preparation. Various physical (e.g., grinding, sieving, diluting) and chemical (e.g., mineralization, extraction, derivatization) processes change the original shape of the sample in order for it to be possible to carry out the measurement for a given property of the test object. In addition, calibration of the measuring instrumentation, carrying out the measurement in adequate working conditions and optimised parameters of the instrument are all important factors. Considering the rules of good laboratory practice, the chemist should perform an optimization of conditions for all steps of the measurement procedure, and then through validation confirm their adequacy for the intended objective of the test. In correctly carried out measurements, relevant standards should be used to enable the traceability of the obtained result. In practice, analytical procedures include measurement of physical quantities; for example, weighing the sample, taking the temperature or diluting the solution in the measuring vessel to a very specific volume. In such cases, traceability is ensured through the use of calibrated tools, including scales and thermometers. Often, the measurement traceability is documented with a relevant calibration certificate, issued by an accredited calibration laboratory. Awareness that ensuring measurement traceability is an indispensable part of the reliable result

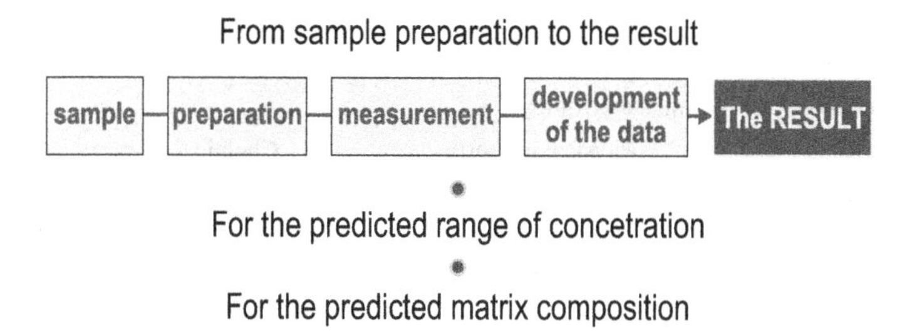

Fig. 1.3 Schemes of the entire measurement procedure in chemical measurements

resulted in the introduction of an international system of units of measurement (SI) and in the signing of the metric convention many years ago. However, it only applies to the measurement of physical properties.

In the case of chemical measurement, it is not possible to set up a system that would ensure measurement traceability for all test objects (samples) and their chemical properties (identification and quantification of the presence of the given compound). This stems mostly from the lack of a complete set of standards (chemical substances) that could be used in all kinds of measurements for all possible measurands. It is also not possible to prepare standards that would reflect the diversity of objects and their matrix. Because of this, in the case of chemical measurements, there is no established infrastructure including calibration laboratories and commonly accepted systems that would ensure traceability. In practice, chemical substances with the highest available purity and well determined chemical content (i.e. primary measurements standard) are used for the calibration of measuring devices, for example, spectrometers, chromatographs and pH meters, termed reference materials (RMs). In order to ensure measurement traceability, while taking into account the influence of the complex composition of the sample on the measurement result, matrix RMs are used, and their goal is to mimic all possible properties of the test object and its behaviour on the stage of sample preparation.

In practice, this means that, for example, during the determination of the fraction of cadmium in the soil—that which is extractable with water at 60 °C—it would be necessary to apply standards for the soil with almost identical granulation and closed content of given cadmium compounds that can be extracted with water at that exact stated temperature. As highlighted above, taking into consideration the variety of objects and the variety of tests carried out, it is not possible to have standards that would fulfil the metrological requirements for all the kinds of tests conducted in chemical laboratories.

A comparison of the most important aspects related to the use of the principles of metrology in physical and chemical measurements has been shown in the Table 1.1.

In recent years, many initiatives have been undertaken in order to introduce the principle of metrology to chemical measurement. Below are examples of initiatives undertaken in the international forum.

Table 1.1 Metrological aspects of the measurements of physical and chemical properties: similarities and differences

Physical metrology vs. chemical metrology

	Physical property	Chemical property
Measurement	Comparison of result of measurement (e.g., temperature)	Comparison of the result of test (e.g., content of DDT in powdered milk)
Units	E.g. m, s, K	E.g. mol/kg, mg/kg
Influencing the result of measurements	Performance of measurement instrument (calibration of instrument)	Type and matrix of examined object; sample preparation procedure (e.g., extraction, digestion); closeness of the RM with the examined object*); performance of measuring instrument (calibration of instrument)
Object of interest	Results DO NOT depend on the type of examined object	Results depend on the type of examined object
Examples	The length of the table; the length of the room	The content of lead in sea water, in soil, in blood

* means the closeness of physical properties (e.g., state of matter, granulation of powder) and chemical properties (e.g., content of analyte and composition of matrix components)

- International Bureau of Weights and Measures (BIPM; *Bureau International des Poids et Mesures*) had brought to life the International Committee for Weights and Measures (CCQM; *Comité Consultatif pour la Quantité de Matière*), whose task is to link the chemical measurements to the international system of units of measurement—SI (www.bipm.fr).
- Eurachem* and CITAC** organizations have prepared a guide for determining the uncertainty of the measurement result in the area of analytical chemistry. All guides are available on their webpages, www.eurachem.org or www.citac.cc, respectively.
- ISO/IEC 17025 standard is a document that holds the requirements regarding the competencies of testing and calibration laboratories, which includes the metrological requirements. This standard is used in the accreditation of test and calibration laboratories.

* Eurachem is a network of organizations in Europe with the objective of establishing a system for the international traceability of chemical measurements and the promotion of good quality practices. It provides a forum for the discussion of common problems and for developing an informed and considered approach to both technical and policy issues.

** CITAC is an organization created in 1993 with the mission to improve traceability of the results of chemical measurement everywhere in the world and to ensure that analytical measurements made in different countries and/or at different times are comparable.

1.3 Metrological Requirements in Chemical Measurements

The evaluation of the value of the measured quantity or—as an analytical chemist would refer to it—the determination of the content of the analyte in the sample usually requires the sampling of a suitable amount of the matter from the representative object delivered to the laboratory. This is followed by the conduct of a whole set of physical and chemical processing enabling, among other things, the isolation of the analyzed substance from the matrix, if necessary its possible concentration and/or change of its chemical form, and ending in the measurement itself for the prepared sample. All the stages involved have a significant influence on the final result of the determination. Evaluating the uncertainty of the final result requires a detailed knowledge of all the steps of the applied measuring procedure, which enforces the need to describe all of its components. According to the principles of metrology, a critical examination of the individual steps of the measuring procedure is one of the more important aspects allowing the evaluation of the quality of the result. It is one of the most important proofs that confirms the competence of the laboratory. The application of the basic principles of metrology to the chemical measurement is not an easy task, most of all because chemical measurements are different from the physical measurements (as discussed earlier). In many cases, it is not possible to directly fulfil all the metrological requirements, hence the need to use the best conduct in a given area, such that it can be and is accepted by all interested parties.

In practice, it means that, for example, if there is a lack of reference material ideally aligned with the test object, carefully selected reference material is used that differs to an acceptable extent. It is allowed and justified only when all laboratories that carry out such tests use the same reference material, traceable to the same standard and the client (e.g., a ministry) is informed about it and accepts such an agreement.

However, the laboratory staff should always do their best to carry out the test/measurements to the best of their knowledge in the area of metrology principles, and try to always follow these principles to the greatest possible extent. In case of the need to accept deviation, staff should have full awareness of their influence on the final result of the measurement procedure.

Two important definitions (ISO/IEC Guide 99)
Quantity: property of a phenomenon, body, or substance, where the property has a magnitude that can be expressed as a number and a reference
A reference can be a measurement unit, a measurement procedure, a reference material, or a combination of such. [Clause 1.1; ISO/IEC Guide 99]
Measurand: quantity to be measured.
The specification of a measurand requires knowledge of the kind of quantity, description of the state of the phenomenon, body, or substance carrying the quantity, including any relevant component, and the chemical entities involved. [Clause 2.3; ISO/IEC Guide 99]

The basic principles of metrology, a specific roadmap that should be in the minds of all those conducting chemical measurements, are as follows:

1. A technologically justified measuring procedure should be selected and then verified (in case of validated methods, the laboratory should confirm the possibility of obtaining the required measuring parameters);
2. The measuring procedure should be described in the form of a mathematical equation (mathematical model);
3. The unit of measurement should be defined, for which the result maintains measurement traceability and demonstrates that traceability (to the greatest possible extent);
4. The uncertainty of the measurement result should be evaluated;
5. The relevant certified reference material (CRM) should be used (to the greatest possible extent).

To conclude, metrology introduces to the area of chemical measurements a unified approach to the evaluation of analytical parameters of a given method. It involves validation and a unified approach to delivering the uncertainty of the measurement result, as well as a unified way to compare the result to the standard—via the traceability.

Chapter 2
Metrological Infrastructure

2.1 A Short History of the Development of International Metrology Infrastructure

1789: Introduction of the metric system
1875: The Metre Convention was signed
1960: Système International d'Unités was established, with the international abbreviation SI
1999: The MRA CIPM [Mutual Recognition Arrangement (MRA); International Committee for Weights and Measures (CIPM; *Comité International des Poids et Mesures*)] was signed.

2.2 The Development of Modern Metrology and International Infrastructure

Nowadays, it is hard to imagine a modern society functioning without an adequate transport and communication infrastructure, as well as without suppliers of goods and services. All of these activities require measurements, and effective operation of related structures would not be possible without an adequate infrastructure to ensure the comparability of measurement results. This means that the metrological infrastructure is of fundamental importance for the development of society. In times of globalization, it must be a structure that transcends borders and political divisions.

Measurements are an integral part of our lives and a tool to describe the world around us through the numbers and the relationships between them. We buy 2 kg of potatoes, 200 g of candy, 30 L of fuel. We pay for consumed water in the house, and the fees are calculated on the basis of the counter (the number of liters of water used). Every day, we check the outside temperature in order to wear suitable clothing; before going on a long journey, we measure the pressure in the tires of our car. We buy

E. Bulska, *Metrology in Chemistry*, Lecture Notes in Chemistry 101,
https://doi.org/10.1007/978-3-319-99206-8_2

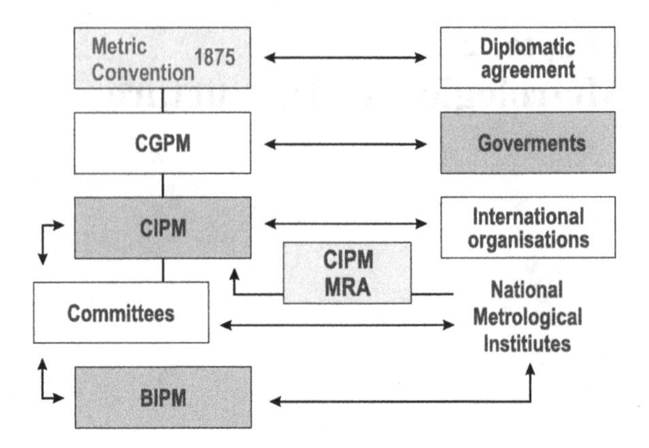

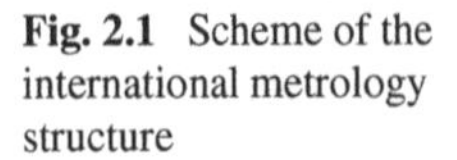
Fig. 2.1 Scheme of the international metrology structure

mineral water containing sodium and potassium ions in expected concentrations. We check the alcohol content in beer and in the wine. Such examples can be multiplied almost indefinitely. And it is important for such actions to be independent of the place and time.

2.3 Metre Convention

The need for a unified system of comparing measurements to a common reference was the driving force behind the creation of a uniform system of measurement, which was the aim of the signatories of the Metre Convention (*Convention du Mètre*), also known as the Treaty of the Metre. The international Treaty of the Metre created the International Bureau of Weights and Measures (BIPM), an intergovernmental organization under the authority of the General Conference on Weights and Measures (CGPM) and the supervision of the International Committee for Weights and Measures (CIPM), which coordinates international metrology infrastructure and the development of the metric system (Fig. 2.1).

The Metre Convention, established in the nineteenth century, was the foundation of the metric system. The ultimate objective of the Metre Convention was to accomplish a state where "regardless of place and time of the measurement, it will yield the same result, within the limits of set errors." Today, this sentence can be finished with the words "within the limits of measurement uncertainty."

In its early stage of development, the metric system of measurement was based on two basic units: meter as the unit of length and kilogram as the unit of mass. This system was adopted by the French National Assembly and, because of its advantages, it quickly gained recognition in the international forum. The consequence of this was the most important event in the history of metrology, namely the signing of the Metre Convention. This event took place on May 20th, 1875, when at a meeting of the Diplomatic Conference in Paris (France), 17 countries signed an agreement on the arrangement of used measurement units. As of 23 March 2018, there are 59 Member States and 42 Associate Member States.

The activities carried out by the signatories of the Metre Convention have led to the uniformity of measures on a global scale, enabling the comparison of results of measurements performed in different places at different times and under different conditions, which is of fundamental importance in science, technology, and the global economy. The important tasks assigned to the Metre Convention, were as follows:

- To define international units of measurement;
- To ensure worldwide uniformity of measurement through the dissemination and improvement of the metric system;
- To ensure equivalence of measurement standards in the signatory countries of the Metre Convention;
- To achieve international compliance measurements, increasing mutual trust in the reliability of measurement results.

In order to accomplish its objectives, the Convention set up three bodies:

- General Conference on Weights and Measures (CGPM; *Conférence Générale des Poids et Mesures*);
- International Committee for Weights and Measures (CIPM; Comité international des Poids et Mesures);
- International Bureau of Weights and Measures (BIPM; *Bureau International des Poids et Mesures*).

The General Conference on Weights and Measures (CGPM) is made up of delegates from all Member States and is convened every four years; it approves the system of SI units of measurement and the results of basic research in metrology. The International Committee for Weights and Measures (CIPM) consists of 18 representatives of the CGPM and supervises the work of the BIPM, appoints the chairmen of Consultative Committees and co-operates with other international metrological organizations. The International Bureau of Weights and Measures (BIPM) is a research institute set up to carry out Convention tasks, acting under the direction and supervision of the International Committee of Weights and Measures. The BIPM makes recommendations regarding the findings in the field of metrology, coordinates research units and standards, and organizes interlaboratory comparisons (called key comparisons) for laboratories maintaining national standards of measurement.

2.4 International Metrological Infrastructure

2.4.1 International Bureau of Weights and Measures (BIPM)

The BIPM is considered to be the world center bringing together the most eminent experts in the field of metrology, which provides the possibility of a global coordination of metrological work at the highest level. BIPM activities primarily include

Table 2.1 List of consultative committees operating under international committee for weights and measures

Name in English	Name in French	Commonly used abbreviation (originating from the French name)
Consultative Committee for Acoustics, Ultrasound and Vibration	Comité Consultatif de l'Acoustique, des Ultrasons at des Vibrations	CCAUV
Consultative Committee for Electricity and Magnetism	Comité Consultatif d'Électricité et Magnétisme	CCEM
Consultative Committee for Length	Comité Consultatif des Longueurs	CCL
Consultative Committee for Mass and Related Quantities	Comité Consultatif pour la Masse et les Grandeurs Apparentés	CCM
Consultative Committee for Photometry and Radiometry	Comité Consultatif de Photométrie et Radiométrie	CCPR
Consultative Committee for Amount of Substance: Metrology in Chemistry and Biology	Comité Consultatif pour la Quantité de Matiére – metrologie en chimie	CCQM
Consultative Committee for Ionizing Radiation	Comité Consultatif des Rayonnements Ionisants	CCRI
Consultative Committee for Thermometry	Comité Consultatif de Thermométrie	CCT
Consultative Committee for Time and Frequency	Comité Consultatif du Temps et des Fréquences	CCTF
Consultative Committee for Units	Comité Consultatif des Unités	CCU

research and the development of international measurement standards, as well as performing calibrations for national metrology institutions (NMIs), coordinating with these institutions and organizing key comparisons.

It is worth mentioning that within the framework of the International Committee for Weights and Measures, dedicated Consultative Committees (CCs) operate dealing with various areas of measurement. The CIPM currently has ten CCs (Table 2.1).

The role of listed above Consultative Committees are crucial, especially in respect to couching the recommendations for CIPM, and in terms of the organization of topic-oriented key comparisons of national measurement standards. Some tasks listed on the BIPM webpage are recall below.

The Consultative Committees have a responsibility:

- To advise the CIPM on all scientific matters that influence metrology, including any BIPM scientific programme activities in the field covered by the CC;
- To establish global compatibility of measurements through promoting traceability to the SI and, where traceability to the SI is not yet feasible, to other internationally agreed references (e.g., hardness scales and reference standards established by the WHO);
- To contribute to the establishment of a globally recognized system of national measurement standards, methods and facilities;
- To contribute to the implementation and maintenance of the CIPM MRA;
- To review and advise the CIPM on the uncertainties of the BIPM's calibration and measurements services as published on the BIPM website;
- To act as a forum for the exchange of information about the activities of the CC members and observers; and
- To create opportunities for collaboration.

www.bipm.org/en/committees/cc/cipm-consultative-committees.html

2.4.2 The International Organization of Legal Metrology (OIML)

The International Organization of Legal Metrology (OIML; *Organisation Internationale de Métrologie Légale*) was formed on October 12th, 1955, in the course of the convention, which took place in Paris (France). This organization includes 58 Member States and 51 Corresponding Members. Its main tasks are: to promote the global harmonization of procedures in the field of legal metrology, making recommendations, documents and dictionaries; studying the issues of metrology in the legally regulated fields; the introduction of the "system of certification of measuring instruments OIML" and the system of recognition for the testing of instrument types.

Legal metrology is the application of legal requirements to measurements and measuring instruments.

The objective of this organization is to create global standards for use in legal metrology legislation and to harmonize the legal metrological requirements for measuring instruments used in the Member States. Within the OIML, the International Bureau of Legal Metrology and the International Committee of Legal Metrology operate. Draft documents of a general nature and recommendations are prepared by the relevant Technical Committees. Different countries use different forms of implementation of OIML documents; for example, in Poland, the native country of the author of this book, the president of the Central Office of Measures introduces these documents in the form of regulations. Besides preparing documents and recommendations, OIML also publishes dictionaries of metrology terms and various publications of a scientific nature.

The project to create the SI system of units was approved at a plenary meeting of the OIML in Paris in 1958. Initially, SI did not include the mole, an extremely important unit for chemistry. Only in 1971 was the mole included as the amount (number) of a substance, a basic unit.

The widespread introduction of the SI system took some time. Taking Poland again as an example, it was only in 1966 that SI was established as a legitimate public system for units of measurement. The Act of June 17th, 1966 on measures and measurement tools (Journal of Laws, 1966, No. 23, item. 148) introduces the guiding principle of the use of the only legal units of measurement, making an exception to this rule only for scientific purposes and for the needs of national defense. Currently, the Act of 11th May 2001 is in force, while in the Journal of Laws, 2004, No. 243, item. 2441 the "Proclamation of Marshall of the Polish Government on November 4th, 2004 on the announcement of the uniform text of the Law Act on Measures" is mentioned.

The world metrological infrastructure for the measurement of physical properties includes BIPM and the network of National Metrology Institutions (NMIs) in the Member States of the Metre Convention. The hierarchy of models providing comparisons of measurement standards of physical quantities is well developed, starting from the international level, through the national level to the calibration laboratories, and consequently to the users of measuring instruments. In this way, the hierarchical system ensures traceability. This allows the user of the measuring instrument to be sure that the measurement results are linked to the national and international standards of measurement units. Of course, this infrastructure is not a guarantee of receiving a reliable result; it is certainly the responsibility of the one who performs the measurements. Therefore, it is necessary to use the validated measurement methods and measuring instruments with the appropriate metrology status, in order to take correct measurements and properly determine the measurement result with its uncertainty.

The signing of the MRA CIPM (Mutual Recognition Arrangement of International Committee for Weights and Measures) by the directors of National Metrological Institutions was an important step in the development and recognition of the uniformity of units at the international level. The Agreement was signed during the 21st General Conference of Weights and Measures, which took place on October 11–15, 1999, as proposed by the International Committee for Weights and Measures. Currently, the MRA is signed by 64 countries, including 45 signatory states of the Metre Convention, 17 associated countries, and two international organizations.

The most important consequence of the signing of the Metre Convention was the creation of National Metrology Institute (NMI) in the signatory countries, and an agreement that the standard units have become units of measurement based on the metric system.

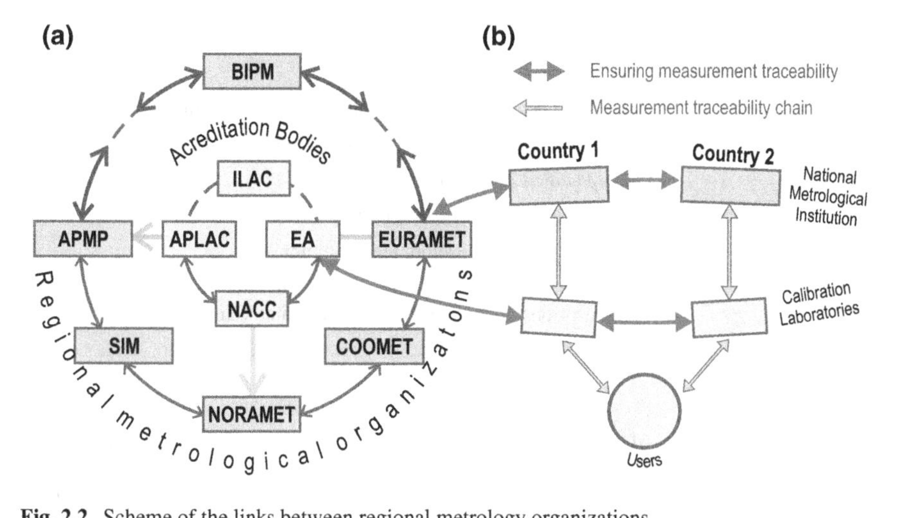

Fig. 2.2 Scheme of the links between regional metrology organizations

The primary responsibility of the NMI in each country is to ensure the uniformity of units and the required accuracy of measurements. Thus, the mission of NMIs is to provide uniformity in the country through the development and maintenance of state measurement standards and the transfer of units of measurement to the market. Each NMI occupies the highest position in the national metrology of economically developed countries: it conducts research activities, performs measurements, consults and acts as a center of expertise metrology. It is a key partner for ensuring the linkage of the national system of measurement with the worldwide system.

2.4.3 Regional Metrological Infrastructure

The activities of metrological organizations on an international scale and their cooperation with national counterparts have been described above. Another level of the infrastructure involves the actively functioning regional organizations that work with both within the Metre Convention and with national institutes. The regional organizations of metrology (e.g., the Regional Metrology Organization; RMO) coordinates activities in the field of metrology on regional level, to disseminate knowledge, organize joint R&D programs, as well as to ensure consistency of measurement of the subordinate NMIs, by organizing key, complementary and bilateral comparisons and through reviewing measurement capabilities declared by the NMIs. Regional organizations are also involved in providing support to developing countries in building and maintaining metrological infrastructure (Fig. 2.2).

WELMEC (European Cooperation in Legal Metrology) operates in the field of legal metrology in Europe. The members of this organization are the national legal

Table 2.2 List of regional metrology organizations

Commonly used name	Full name of the organization
EURAMET	The European Association of National Metrology Institutes
COOMET	Euro-Asian Cooperation of National Metrological Institutions
APLMF	Asia-Pacific Legal Metrology Forum
APMP	Asia Pacific Metrology Programme
SADC	Southern African Development Community
SIM	Inter-American Metrology System with five sub-regional organizations, including: Noramet, Carimet, Camet, Andimet, and Suramet

metrology authorities of the Member States of the European Union and members of the European Free Trade Association (EFTA). The main goal of the WELMEC organization is to promote mutual confidence in legal metrology between member countries, harmonization of activities in the field of legal metrology and to promote the exchange of information between the concerned institutions. The principal aim of WELMEC is to establish a harmonized and consistent approach to European legal metrology. Originally, WELMAC (original name: Western European Legal Metrology Cooperation) was established by 13 European countries—the Memorandum of Understanding was signed in June 1990, in Bern (Switzerland). Currently, 39 countries (31 regular and 8 associate members) are represented in the WELMEC Committee. WELMEC has also signed Memorandums of Understanding with the EA (European co-operation for Accreditation) and EURAMET (European Collaboration in Measurement Standards).

EURAMET is a regional organization focusing on collaboration in measurement standards. The organization was established on 23 September 1987 as a facultative forum for the international cooperation of European National Metrological Institutions in the implementation, maintenance and reproduce of national measurement standards. The most important tasks of the EURAMET include the creation of a framework for the cooperation of national metrology institutes, organizing comparisons and research projects, the transfer of experience related to the construction and maintenance of national measurement standards, collaboration with calibrating laboratories and legal metrology. Apart from the European organizations, there are also many other regional organizations in the world, which are summarized in Table 2.2.

2.4.4 *The National Metrological Infrastructure*

In every civilized country, the development of the economy and services would not be possible without maintaining an appropriate system of metrology; this is also true for ensuring the health and safety of citizens. Within this, the most important systems are the National Metrological Institutions, the National Accreditation Bodies and the National Institute for Standardization. In addition, the network of calibration laboratories carrying out the relevant work in terms of maintaining the metrological status of the instruments, standards, and measurements is also necessary.

The National Metrology Institute (NMI) in the country, is usually appointed by decision of the competent authority of the state, and its task is to ensure the uniformity of measurement in the country through the development and maintenance of national measurement standards and the transfer of these units. The next level in the hierarchy is the calibration laboratories, ensuring the highest level of accuracy of measurements in the country, comparable to the primary laboratories operating at the NMI. Accredited calibration laboratories whose technical competence, quality and impartiality have been confirmed by an Accreditation Body in the country play a crucial role in this system.

The NMI is an essential partner in the global metrological infrastructure and occupies the highest position in the national metrology of economically developed countries. Its main task is to link the national system with the worldwide measurement system.

2.4.5 Summary

The main aim of building and maintaining the metrological infrastructure on a global level is to sustain the highest standards in respect to traceability, accuracy and reliability of measurements used in industry, transport, communication, trade, medicine and health protection, as well as in many other disciplines in science and engineering. This aim is realized chiefly by multilateral collaboration on international, regional as well as national levels in the area of metrology (Table 2.3).

This aim encompasses several commonly accepted tasks:

- Acceptance the of international system of units, SI;
- Establishing and maintaining standards of units;
- Collaboration with NMIs;
- Accreditation of testing and calibration laboratories;
- Activity of standardization bodies;
- Enforcing the common implementation of legal metrology.

Table 2.3 Areas of collaboration between regional and international metrology organizations

Areas of collaboration	Organization involved in cooperation	
	International	Regional
International system of units SI	BIPM	–
Collaboration of NMIs	BIPM	APMP (Asia and Pacific) COOMET (Europe and Asia) EURAMET (Europe) SIM (American countries) SADCMET (South America) NORAMET (USA, Canada, Mexico)
Accreditation	ILAC	APLAC (Asia and Pacific) EA (Europe) IAAC (American countries) NACC (Nord America) SARAC (South Africa)
Legal metrology	OILM	APLMF (Asia and Pacific) COOMEF (Europe and Asia) SALMEC (South Africa) WELMAC (Europe)
Standardization	ISO/IEC	AIDMO (Arabic countries) ACCSQ (Middle East) ARSO (Africa) CEN/CENELEC (Europe) COPANT (American countries) PASC (Region of Pacific) SADSTAN (South Africa)

BIPM: International Bureau of Weights and Measured; ILAC: International Laboratory Accreditation Cooperation; ISO: International Organisation for Standardization; IEC: International Electrotechnical Commission; OILM: International Organisation of Legal Metrology

Chapter 3
System of Units

3.1 The Value of Defined Quantity Is the Quantity Expressed as the Product of Number and Unit of Measurement

Information about the environment is obtained through the measurement of various quantities, for example, physical and chemical. These quantities allow characterization of the properties of objects (samples), phenomena or processes. Such properties can be determined qualitatively and quantitatively. They can also be compared qualitatively and quantitatively with the same properties of other objects or phenomena. The term 'quantity' means the amount in a general sense (length, time, mass, concentration) or a specific value (length of the rod, the weight of the weight, the journey time, the concentration of sodium ions in the mineral water). Measurement as such is an experimental activity, aiming to determine the value of the quantity expressed as a number and unit of measurement.

Measurement unit: real scalar quantity, defined and adopted by convention, with which any other quantity of the same kind can be compared to express the ratio of the two quantities as a number. [clause 1.9; ISO/IEC Guide 99]
System of units: a set of base units and derived units, together with their multiples and submultiples, defined in accordance with given rules, for a given system of quantities.
[clause 1.13; ISO/IEC Guide 99]

Hence, metrology can be considered as a scientific discipline dealing with setting the dimensions of the measured quantities and their units as well as methods of measuring.

Quantities used to characterize the properties of the objects are divided into base and derived quantities. **Base quantities** are those that in a given unit system are

E. Bulska, *Metrology in Chemistry*, Lecture Notes in Chemistry 101,
https://doi.org/10.1007/978-3-319-99206-8_3

assumed to be independent of each other and that can be used, using the formulas, to express the derived quantities of the system. Therefore, the **derived quantity** is a quantity determined in a given system as a function of the basic quantity of the system. From the above description, it can be seen that the appropriate system of quantities is an important aspect in ordering the units of measurement.

Apart of the system of quantities, it is also necessary to establish a system of units of measurement that is an ordered set of units of measurement of the base and derived units related to a specific system of quantities.

The unit of measurement of a given quantity is determined in order to allow quantitative comparison of the various values of the same quantity. The symbol of a unit of measurement is called a contractual symbol indicating the unit of measurement; for example, m, a symbol for meter; kg, a symbol for kilogram; and s, a symbol for second. Units of measurement, as well as physical quantities, are divided into the base and derived units of measurement. The system of units thus consists of primary units of measurement, assumed contractually, and the derived units built on them, of a complex nature. Some derived units have names and symbols of their own; for example, newton (N), volt (V).

The ***base unit of measurement*** is the unit of measurement of individual basic quantities.
The ***derived unit of measurement*** is the unit of measurement of one of the derived quantities.

Establishing the systems of units was preceded by establishing adequate reference units of various quantities. These standards, taken from nature or assumed contractually, made it possible to take measurements (see Chap. 2). Through the combination of the chosen units of other quantities as derivatives were formed. For each quantity, any value can be basically assumed as its unit.

As already mentioned, the idea of the metric system, the system of units based on the meter and kilogram, was created in France, when two platinum reference standards artefacts for the meter and the kilogram were deposited in the French Archives de la République in Paris in 1799. The metric system uses decimal multiples and submultiples of units of measurement. Later on, the French Academy of Sciences was obliged by the National Assembly to design a new system of units for the world to use, and as a result, in 1946, the MKSA (meter, kilogram, second, ampere) system was accepted by the countries of the Metre Convention. In 1954, the MKSA system was extended to incorporate the inclusion of the kelvin and the candela, and the adopted the name of the international system of units—SI. This system was established in 1960 by the 11th General Conference of Weights and Measures (CGPM).

Another system of units, ensuring their consistency with the base units of the system, was the CGS (centimeter, gram, second) system. It was initiated by C.F. Gauss, who proposed in 1832 the principle of the structure of the system of units, consisting of choosing the base quantities and establishing their units, which are also used to make other derived units. Gauss was the first to make absolute measurements of the Earth's magnetic fields in terms of a decimal system. In the system proposed by Gauss, units of all the physical quantities are expressed by a combination of three base units: length, mass and time. Gauss called this system, the system of absolute units. Initially, the system consisted of three base units: a millimeter, a milligram and a second. Later, the system was based on the centimeter, gram and second. In the following years, various changes were introduced and the system was expanded by adding kelvin (K) as the unit of thermodynamic temperature, mole (mol) as the unit of amount of matter and candela (Cd) as the unit of light.

During the development of a systems of units, two parallel trends were visible: a 3D CGS system based on three units (base quantities): a centimeter, a gram, a second; and a 4D system MKSA, which arose as a result of the development of the metric system.

The metric system of units of measurement is older than the CGS system: the former was originally designed for the measurement of length, area, volume and weight, which were based on two units: length in meters (initially adopted as one ten-millionth part of one-quarter of the meridian), and mass in kilograms (defined initially as the mass of one cubic decimeter of distilled water at its maximum density). The metric system was adopted in 1791 by the French National Assembly convened in order to establish uniformity of measurement: in the second half of the nineteenth century, it has become the international system. At the Diplomatic Treaty of 17 countries on May 20, 1875, the Metre Convention was created, whose task was to improve international and state standards (etalons) of individual units of measurement and organize the system of units in use. The focus of the Metre Convention was the decimal metric system based on prototypes of meter and kilogram, as well as on decimal division.

The BIPM has worked on the further development of the MKSA enforced in countries that have signed the Metre Convention. At one of the sessions of the International Committee of Weights and Measures, a Commission emerged whose task was to develop MKSA. The Commission proposed a system with an enlarged—compared with MKSA—number of base units, which also included: a unit of temperature at absolute thermodynamic scale (kelvin) and the unit of luminous intensity (candela). This system was abbreviated as SI.

The BIPM approved the development of the Commission and submitted a draft of the SI system to the International Organization of Legal Metrology (OIML), which approved the project at a plenary meeting in Paris on October 7, 1958. In 1971, the unit of the amount (number) of matter (mole) and derived units containing a mole were included as base units and, therefore, became the SI units.

A major achievement of French science was the introduction for the first time of the metric system based on standards of length (meter) and weight (kilogram), using the principle of 10-fold multiples when creating secondary units. Incidentally, the artefact realization of these secondary units was created, considered today as outdated. The model of a meter is a platinum–iridium rod with a rectangular cross-section, the length of which was designated as the distance between the leading planes of the rod, and was one ten-millionth of the distance of a quarter of the Earth's meridian passing through

Fig. 3.1 Model of weight standard (made by Główny Urząd Miar in Warsaw, Polish NMI)

Paris, of course. Almost 100 years later, in 1875, the Metre Convention decided to establish the International Bureau of Weights and Measures (BIPM), whose first task was to physically realize and store prototypes of both meter and kilogram. As part of the introduction of the international system of units (SI), a standard for the metric unit of length (meter) became a platinum–iridium rod with a cross-section having a shape of X, made in 1889. The intention was to prepare a standard of length as close as possible to the length of the archival standard. In subsequent years, the development of laser technology and the possibility to measure optical frequencies allowed the unit of length to be redefined. Now, the meter is defined by the fixed numerical value of the speed of light in vacuum c to be 299,792,458 when expressed in the unit m s^{-1}, where the second is defined in terms of the cesium frequency $\Delta\nu_{Cs}$.

The mass standard (kilogram) corresponded originally to the mass of one cubic decimeter of distilled water at 4 °C (Fig. 3.1). Currently, it is defined by taking the fixed numerical value of the Planck constant h to be $6.626070040 \times 10^{-34}$ when expressed in the unit J s, which is equal to kg m^2 s^{-1}, where the metre and the second are defined in terms of c and $\Delta\nu_{Cs}$

In the previous definition of the kilogram, the value of the mass of the prototype of the kilogram was fixed at one kilogram exactly, and then the value of the Planck constant h had to be determined experimentally. Currently, the definition requires the h value to be exactly fixed; thus, the mass of the prototype has to be determined experimentally.

As previously highlighted, the International System of Units (SI) is a consistent set of units adopted and recommended by the General Conference of Weights and Measures. It is designed to be used in all aspects of human activities, trade, transport, communication, engineering, industry, security, medicine, health and environment safety—to mention just the most important areas. Currently, it consists of seven base units, which together with the derived units make up a coherent system of units of measurement. In addition, certain entities outside of the system are accepted for use with SI units.

The **base units** of the international system of units (SI):

- **The second** (s) is the unit of time, defined by fixed numerical value of the caesium[1] frequency $\Delta \nu_{Cs}$, the unperturbed ground-state hyperfine transition frequency of the caesium 133 atom, at 9,192,631,770 when expressed in the unit Hz, which is equal to s^{-1}.
- **The meter** (m) is the unit of length, defined by the fixed numerical value of the speed of light in vacuum, c, at 299,792,458 when expressed in the unit m s^{-1}, where the second is defined in terms of the caesium frequency $\Delta \nu_{Cs}$.
- **The kilogram** (kg) is the unit of mass, defined by taking the fixed numerical value of the Planck constant h to be $6.626070040 \times 10^{-34}$ when expressed in the unit J s, which is equal to kg m^2 s^{-1}, where the metre and the second are defined in terms of c and $\Delta \nu_{Cs}$
- **The ampere** (A) is the unit of electric current, defined by fixed numerical value of the elementary charge e at $1.6021766208 \times 10^{-19}$ when expressed in the unit C, which is equal to A s, where the second is in terms of $\Delta \nu_{Cs}$.
- **The kelvin** (K) is the unit of the thermodynamic temperature, defined by taking the fixed numerical value of the Boltzmann constant k at $1.38064852 \times 10^{-23}$ when expressed in the unit J K^{-1}, which is equal to kg m^2 s^{-1} K^{-1}, where the kilogram, metre and second are defined in terms of h, c and $\Delta \nu_{Cs}$.
- **The mole** (mol) is the unit of amount of substance of a specific elementary entity, which may be an atom, molecule, ion, electron or any other particle or a specified group of such particles, defined by the fixed numerical value of the Avogadro constant N_A at 6.022408 s57 $\times 10^{23}$ when expressed in the unit mol^{-1}.
- **The candela** (cd) is the unit of luminous intensity in a given direction, defined by taking the fixed numerical value of the luminous efficacy of monochromatic radiation of frequency 540×10^{12} Hz, K_{cd}, at 683 when expressed in the unit lm W^{-1}, which is equal to cd sr W^{-1}, or cd sr kg^{-1} m^{-2} s^3, where the kilogram, metre and second are defined in terms of h, c and $\Delta \nu_{Cs}$.

Derived unit is a unit of measurement of the derived quantity in a given system of quantities. Derived units in SI are linked to the base SI units, according to the physical dependencies between quantities. The derived unit is, among others: the unit of area in square meters (m^2); volume unit in cubic meters (m^3); and linear speed in meters per second ($m \bullet s^{-1}$). Some of them are given special names (Table 3.1).

SI units of measurement are recommended in all cases of international trade and the regulations are adopted and accepted almost all over the world. However, there are many units that do not belong to the international SI system, but they

[1]The element **Cs** was named after the Latin word *caesius*, meaning bluish grey. *Caesium* is the spelling recommended by the International Union of Pure and Applied Chemistry (IUPAC); the American Chemical Society (ACS) has used the spelling *cesium* since 1921, following Webster's New International Dictionary.

Table 3.1 List of selected examples of derived units

Quantity	Unit	Symbol	Reference to SI units
Plane angle	Radian	rad	rad = m/m
Solid angle	Rteradian	sr	sr = m^2/m^2
Frequency	Hertz	Hz	Hz = s^{-1}
Force	Newton	N	N = kg m s^{-2}
Electric charge	Coulomb	C	C = A s
Electric conductance	Siemens	S	S = kg^{-1} m^{-2} s^3 A^2
Temperature in Celsius	Degrees Celsius	°C	°C = K
Activity referred to a radionuclide	Becquerel	Bq	Bq = s^{-1}

Table 3.2 Units outside of the SI (non-SI units) accepted for use with the SI units

Quantity	Unit	Symbol	Reference to SI units
Time	Minute	min	1 min = 60 s
Time	Hour	h	1 h = 3600 s
Time	Day	d	1 d = 86,400 s
Volume	Litre	l, L	1 L = 10^{-3} m^3
Mass	Tonne	T	1 t = 103 kg
Area	Hectare	ha	1 ha = 10^4 m^2

are nonetheless widely used and are part of the culture and civilization. Therefore, in 1966, the International Committee of Weights and Measures (CIPM) accepted those units whose practical application is justified. Selected examples are shown in Table 3.2.

It worth noting that the creation of the units reflects in many cases the development of science, and several units have received names derived from the names of distinguished scientists who have worked in the field. The list of these individuals and the list of related names is long. Often, people whose names are the root of the name of the unit—physicists, mathematicians and chemists—are from different countries. Several examples are given in Table 3.3.

Units of measurement are implemented in the form of individual standards. Standards of the highest metrological hierarchy are kept at the International Bureau of Weights and Measures and are consistent with their national standards maintained by the National Metrological Institutes.

The obligation to apply the legal units of measurement concerns the use of measuring instruments, measuring the expression of the values of physical quantities in the economy, health care and public safety and the activities of an administrative nature.

Table 3.3 Units with names of distinguished scientists

Name of unit	Name of scientist	Country of origin
Ampere	André Marie Ampère	France
Hertz	Hendrich Hertz	Germany
Newton	Sir Isaac Newton	England
Ohm	Simon George Ohm	Germany
Pascal	Blaise Pascal	France
Sievert	Rolf Sievert	Sweden
Tesla	Nikola Tesla	Croatia
Volt	Alessandro Volta	Italy
Weber	Wilhelm Eduard Weber	Germany

3.2 Conventions for Writing Unit Symbols and Their Names

Awareness of the rules of the writing and the names of SI units is a prerequisite for their use. It is worth remembering that the letters are used both as symbols of physical quantities and units of measurement. For different purposes, we use simple, slanted or bold letters. The question is how to write numbers, symbols of physical quantities, unit symbols, and symbols of elements: which should be written in a simple (Roman) style, which should be slanted (in italics) and which should be written in bold.

The commonly used rules and those accepted by consensus are straightforward: the symbols of units are written in lowercase letters (e.g., m, s, kg), and only the names of units that are derived from a name are written as a capital letter; for example, A, Wb, Hz. The exception relates to the liter, which allows the use of two symbols (l or L). In order to avoid possible confusion between the numeral 1 (one) and the lowercase l (el), the symbol L is recommended. The very name of the unit is always lowercase: for example, newton (N), second (s); the exception is the Celsius degree.

The unit symbol is not an acronym, but rather the mathematical quantity and is usually marked with the first letter of the name of the unit (s is the symbol of a second, not sec; g is the symbol of a gram, not gm) with some exceptions, such as Cd, mole, Hz. In the case of the temperature, expressed in kelvin (name of the unit of thermodynamic temperature), we use the symbol K (note a Kalven degree or °K). They remain, however, in the unit of Celsius temperature. It is, therefore, Celsius degree, °C. The symbols of units and their prefixes are written in the Latin alphabet, with the exception of the symbols for the ohm (Ω) unit and the prefix micro (μ), which is written in the Greek alphabet.

All symbols of physical quantities are written in Times New Roman font, italic, regardless of the type of font used throughout the document. This principle is extremely important because its non-compliance can lead to confusion of unit symbols with a symbol of the measured value. For example, the symbol of the mass (*m*), and the unit symbol for length in meters (m).

Table 3.4 Basic roules for writing symbols and unit's name

The symbols and unit names should not be combined in the record	
Correct	Incorrect
10 g/kg or 10 g • kg^{-1}	10 g per kilogram
10 grams per kilogram	10 grams per kg

Table 3.5 Basic roules for writing units with a numerical values

Assigning units to a numerical value should be unequivocal	
Correct	Incorrect
35 g × 48 g	35 × 48 g
100 g ± 2 g	100 ± 2 g
(100 ± 2) g	

- The quantity of the measured **mass** is marked with the symbol (m), a unit of mass is the kilogram, and a symbol is a unit of mass (kg).
- The quantity of the measured **length** is marked with the symbol (l), the unit of length is the meter, a unit symbol of length is (m). The symbols and unit names should not be combined (Table 3.4).

3.3 Editorial Requirements with Regard to Writing SI Units

- The designation of the symbol of unit remains unchanged in the plural (we do not add the plural suffix);
- After designating the unit (symbol) we do not add a full stop to indicate a contraction; unless it is the end of the sentence;
- Complex units, created by the multiplication of several units must be written as follows: e.g., Nm, N m, N • m, N × m;
- Complex units, created by dividing of several units must be written as follows: e.g., m/s or m s^{-1};
- Designations of units must be separated by a space from the numeric value, e.g., 8 kg and not 8 kg;
- Assigning units to a numerical value should be unequivocal (see Table 3.5).

The prefix symbol is placed before the symbol of the unit, without a space; for example, μg, not μ g.

Table 3.6 Basic roules for writing symbol of the unit of volume - liter

Examples of recording units of liter		
Type of character	Lowercase	Uppercase
Times New Roman	0.001 l	0.001 L
Arial	0.001 l	0.001 L

3.4 Writing the Unit Symbol of Volume

The common principle is that the symbols of units are written in a lowercase letter; thus, the recognized symbol of a liter is the lowercase letter 'l'. The prevalence of computers meant that such a rule is often confusing (owing to the potential confusion with the number 1), so it was accepted that the liter unit could be designated with a capital L (Table 3.6).

3.5 Quantities, Units of Measurement and Related Concepts

The measurable quantity: a feature of the phenomenon, body or substance that may be distinguished qualitatively (from other properties) and determined quantitatively.

The system of quantities: the set of quantities, between which there are certain relationships, that contain a specific group of the base and derived quantities.

Base quantity: an arbitrarily chosen quantity in a given system of quantities, which is an elementary notion—that is, a notion not requiring determination by means of other quantities, independent of the others, such as mass, length and time.

Derived quantity: the quantity defined, in a specific system of quantities, as a function of the underlying base quantities of the system; for example, speed.

The dimension of the physical quantity: an expression that represents the physical quantity of the given system of quantities as a product of powers of factors indicating the base quantities of the system with a numerical factor equal to one.

The value of the physical quantity: the quantitative expression of a quantity, generally in the form of the product of number and unit of measurement.

Unit of measurement of quantity: the quantity specified, defined and adopted by the convention, which is compared to other quantities of the same kind for the purpose of quantitative expression in relation to the quantity adopted by the convention. Such a quantity is assigned a numeric value equal to one.

The symbol of a unit of measurement: a contractual sign indicating the unit of measurement (e.g., m, meter; A, ampere). Symbols of units of measurement may be single: m, meter; K, Kelvin; or they may be complex: $m \cdot s^{-1}$, meter per second; kg/L, kilogram per liter.

The base unit of measurement: the unit of measurement of the base quantity in a given system of quantities.

The derived unit of measurement: the unit of measurement of the derived quantity in a given system of quantities.

Consistent unit of measurement: this may be expressed as the product of powers of base units with a coefficient of proportionality equal to one; for example, in SI, the consistent unit w is 1 N = 1 m kg s^{-2}.

Unit of measurement outside of the system: this does not belong to the system of units, for example, day, hour, minute are units of time outside the SI.

Dimensionless unit of measurement: the derived unit of measurement with a dimension of one; for example, a unit of plane angle (radian; rad) or solid angle (steradian; sr).

Legal unit of measurement: the unit of measurement, the application of which is required or permitted by legislation.

The system of units: an ordered set of units of measurement, created on the basis of the conventionally adopted base quantities and with assigned units of measurement and set equations used to define the derived quantities.

A coherent system of units of measurement: derived units of measurement are expressed by base units with a formula with a numerical coefficient of one.

The equation of units: specifies the units for derived quantities in a certain system of units through base and other derived units of the said system; for example, 1 m^2 = 1 m · 1 m, 1 W = 1 V · 1 A. Also establishes the relationship between the unit of measurement of a certain size and its multiple or sub-multiple; for example, 1 mm = 0.001 m, 1 pF = 10^{-12} F. Determines the equivalence between the measurement units of the same quantity in the different systems of units; for example, 1 kg = 1 kg · 9.80665 m/s^2 = 9.80665 N.

3.6 Fundamental Physical Constant

The fundamental physical constants act as universal coefficients binding certain quantities. They are present in the equations expressing the laws of nature in the form of products of powers of base quantities.

Characteristics of constants:

- Generally believed to be both universal in nature and having constant value in time;
- It is unlike a mathematical constant, which has a fixed numerical value, but does not directly involve any physical measurement;
- They have assigned units of measurement;
- Their values do not change in the adopted system of units of measurement;
- In the quantitative equations, they are treated as individual quantities;
- Immutable parameters of the universe influence their actual form.

The fundamental values of the constants are recommended by the Committee on Data for Science and Technology (CODATA). The CODATA Task Group on Fundamental Physical Constants has been publishing recommended values for the Fundamental Physical Constants since 1969. The most updated information can

Table 3.7 Fundamental physical constants

Fundamental constant	Symbol	Value	Unit	Uncertainty
Speed of light in vacuum	c	299,792,458	$m \cdot s^{-1}$	Exact
Magnetic constant (vacuum permeability)	μ_0	$4\pi \times 10^{-7}$ $N \cdot A^{-2}$ = $1.256637061\ldots \times 10^{-6}$	$N \cdot A^{-2}$	Exact
Electric constant (vacuum permittivity)	ε_0	$8.854187817\ldots \times 10^{-12}$	$F \cdot m^{-1}$	Exact
Newtonian constant of gravitation	G	$6.67408(31) \times 10^{-11}$	$m^3 \cdot kg^{-1} \cdot s^{-2}$	$\pm 0.00170 \times 10^{-11}$
Planck constant	h	$6.626070040(81) \times 10^{-34}$	$J \cdot s$	$\pm 0.0000080 \times 10^{-34}$
Avogadro constant	N_A	$6.022\ 140\ 857(74) \times 10^{23}$	mol^{-1}	$\pm 0.0000072 \times 10^{23}$
Boltzmann constant	k	$1.38064852(79) \times 10^{-23}$	$J \cdot K^{-1}$	$\pm 0.0000024 \times 10^{-23}$

be found in *Reviews of Modern Physics,* vol. 84 (2012), written by Peter J. Mohr, Barry N. Taylor, and David B. Newell from the National Institute of Standards and Technology, USA (Table 3.7).

In 1997, by the efforts of the BIPM, the Joint Committee for Guides in Metrology (JCGM) was established, whose task was, among other things, drafting the dictionary aiming to incorporate vocabulary of basic and general terms in metrology, named VIM (fr. Vocabulaire International de Métrologie). VIM was the very first dictionary of metrology and was issued in 1984.

The third edition of the dictionary (VIM 3), with updated title "International Vocabulary of Metrology—Basic and General Concepts and Associated Terms (VIM)," was released as an ISO Guide in 2007. Further work of the working group led to the second edition of ISO/IEC Guide 99:2012 (VIM 3), which is available on the home page of the BIPM (www.bipm.org/vim). This edition is designated JCGM 200:2012.

Box 3.1
VIM: ed. 1984
VIM 2: ed. 1993
VIM 3: ed. 2007
VIM 3, updated: ed. 2012

Measurement standard (etalon): realization of the definition of a given quantity, with a stated quantity value and associated measurement uncertainty, used as a reference.

International measurement standard: measurement standard recognized by signatories to an international agreement and intended to be used worldwide.

National measurement standard: measurement standard recognized by national authority to serve in a state or economy as the basis for assigning quantity values to other measurement standards for the kind of a quantity concerned.

Primary measurement standard: measuring standard established using the primary reference measurement procedure, or created as an artifact, chosen by convention.

Secondary measurement standard: measurement standard established through calibration with respect to a primary measurement standard for the quantity of the same kind.

Reference measurement standard: measurement standard designed for the calibration of other measurement standards of the quantities of a given kind in a given organization or at a given location.

Working measurement standard: measurement standard that is used routinely to calibrate or verify measuring instruments or measuring systems.

Note 1: A working measurement standard is usually calibrated with respect to a reference measurement standard.

Note 2: In relation to verification, the term 'check standard' or 'control standard' are also sometimes used.

There are several organization on international or regional level aiming to support the implementation of metrology system. The list of most relevant is given in Table 3.8.

Table 3.8 List of relevant international organizations

Used name	Full name
BIPM	Bureau International des Poids et Mesures
OIML	Organisation Internationale ds Métrologie Légale
ISO*	International Organization for Standardization*
IEC	International Electrotechnical Commission
IUPAP	International Union of Pure and Applied Physics
IUPAC	International Union of Pure and Applied chemistry
ICTNS	International Committee on Terminology, Nomenclature and Symbols
WELMEC	European Cooperation in Legal Metrology
EURAMET	European Collaboration in Measurement Standards

*ISO is not an abbreviation, it is rather the easy-to-pronounce, accepted name of the organization [otherwise it could be the International Organization for Standardization (IOS) or, in French, *Organisation Internationale de Normalisation* (OIN)]

Chapter 4
Metrological Traceability

Traceability: tight connectivity, compactness. Traceability is a series of logical cause-effect relationships.

Traceability is the capability to trace something, to verify the history, location, or application of an item by means of documented recorded identification.

The measurement is a process in which the unknown quantity is compared with the known quantity, namely with a measurement standard. As discussed in the previous chapters, using the standard of the highest order is not always possible or economically justified. Thus, secondary standards can be used (standards maintained in an accredited calibration laboratory) and compared within the chain of traceability with the standard of the highest order (primary standard). According to the principles of metrology, a suitable standard should be used for measuring the property to demonstrate traceability of the result to the highest possible standard in the given conditions of metrological quality. Accordingly, an international system of units (SI) was introduced for the measurement of physical quantities, and the Metre Convention was signed. The definition of the metrological term 'traceability' has undergone change in the years since the signing of the Metre Convention. In the second edition of the VIM 2 dictionary (1993), an important feature of the definition was a reference to international units of measurement, or to national standards. In the current issue of VIM 3 (2012 r.), the definition emphasizes the relationship of the measurement result with a specific reference, but without the strict requirement that it is the standard of national or international units of measurement.

Metrological traceability—property of a measurement result whereby the result can be related to a reference through a documented unbroken chain of calibrations, each contributing to the measurement uncertainty.
Clause 2.41; ISO/IEC Guide 99

E. Bulska, *Metrology in Chemistry*, Lecture Notes in Chemistry 101,
https://doi.org/10.1007/978-3-319-99206-8_4

A novelty in the current edition of the VIM 3 dictionary is that the definition of traceability is supplemented with eight comments, in which additional explanations are given. In the eighth comment, the recommendation is underlined that, in relation to the metrological use, the full wording of the term 'measurement traceability' should be used, since the term 'traceability' is also used in other contexts, whereas the word 'traceable' refers to the history of the object. Table 4.1 cites a few selected comments that directly relate to the definition.

Metrological traceability, defines the relationship of the measurement result to the accepted reference systems through an unbroken chain of comparisons. The definition of metrological traceability, as explained in Note 2, indicates that traceability requires a hierarchical set of calibrations. The structure of it may be different depending on the purpose of the measurement, but the lowest level is always the result of the particular measurement, and the highest level is the standard of the highest, available in that field, metrological quality.

Figure 4.1 shows a model diagram of the hierarchy of calibration, providing both measurement traceability, as well as the comparability of the results of measurements performed in the laboratory (LAB1 to LAB4).

Metrological traceability of the measurements to the respective standards held in the NMIs, which are traceable to standards held at the International Bureau of Weights and Measures (BIPM), provides the ability to compare the results of all related parties in the traceability chain (the laboratories). Metrological traceability implemented as shown in Fig. 4.1 indicates that the measurement result, regardless of where it was made, can be associated with the state or international (BIPM) standard unit of measurement.

Table 4.1 Selected comments (notes) relating to the definition of traceability

Note	Description
Note 1	For this definition, a 'reference' can be a definition of a measurement unit through its practical realization, or a measurement procedure including the measurement unit for a non-ordinal* quantity, or a measurement standard
Note 2	Metrological traceability requires an established calibration hierarchy
Note 3	Specification of the reference must include the time at which this reference was used in establishing the calibration hierarchy, along with any other relevant metrological information about the reference, such as when the first calibration in the calibration hierarchy was performed

*Ordinal quantity, defined by a conventional measurement procedure, for which a total ordering relation can be established, according to magnitude, where other quantities of the same kind, but for which no algebraic operation among those quantities, exist
(Definition of Traceability in Clause 1.26; ISO/IEC 99)

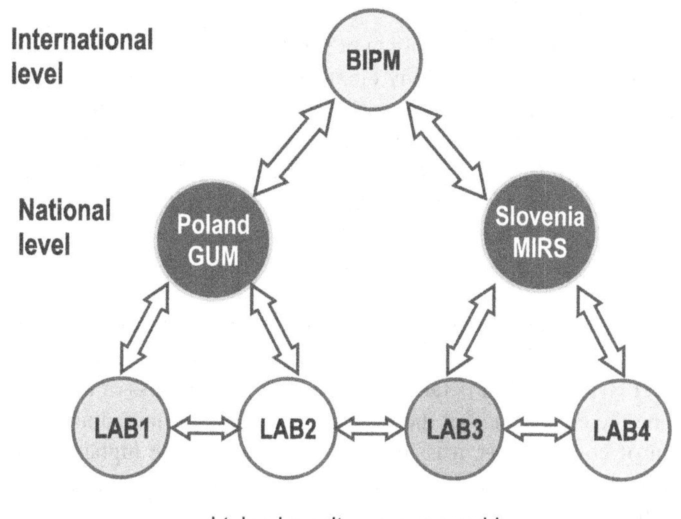

Fig. 4.1 Comparability of results via the metrological traceability to SI units. BIPM: International Bureau of Weights and Measures, GUM (pl., Główny Urząd Miar): Polish NMI, LAB: Given laboratory

4.1 Traceability and the Comparability of Results

A meaningful comparison of the results is valid when they are expressed in the same units or refer to the same scale of measurements. Most of the currently used units of measurement belong to the international system of units (SI) (m, kg, s, A, K, moles, cd). In the case of measurements, where the result is expressed in these units, we assume the same consistency of values of quantity as the SI unit. In the case of metrological traceability in chemical measurements, it is the relation of the result to the standard unit of measurement—the mole. Unfortunately, the mole is the only SI unit that has not been represented in the form of a specific, single standard.

This is mainly due to the huge variety of chemical individuals (atoms, ions, molecules, and others), for which the appropriate standards should be prepared. Ideally, it would be expected to have a standard mole of each of the individual species; for example, of calcium carbonate, or selenium atoms. However, with the current technology this is not possible, and it probably will not be possible in the near future to produce a substance consisting solely of calcium carbonate particles or of selenium atoms. We must also remember that many elements are found in nature in the form of several isotopes with different atomic masses. For example, calcium is present in the form of six stable isotopes of different atomic weights: 40, 42, 43, 44, 46 and 48 a.m.u. This means that for each isotope of calcium, an appropriate standard compound should be prepared. The variety of chemical substances and the constrained possibility of obtaining them in a form that contains exclusively the given atoms or molecules, causes the mole standard to be a virtual unit.

The mole (SI base unit for the amount of substance) is the amount of substance of a system that contains as many elementary entities as there are atoms in 0.012 kg of carbon-12. When the mole is used, the elementary entities must be specified and may be atoms, molecules, ions, electrons, other particles, or specified groups of such particles.

IUPAC. Compendium of Chemical Terminology, (2014) 2nd ed. (the 'Gold Book').

Metrological traceability can be defined not only by comparison with the standard unit of measurement of SI but also in relation to the generally accepted scale of values; for example, pH, hardness or octane scale. However, regardless of the way in which we determine the consistency of the measurement result obtained and the structure of the chain of comparisons, it is important to agree on the highest measurement standard of the hierarchy. The lack of a hierarchical system of comparisons leads to possible multiplication of systematic error (e.g.,.bias).

One particular anecdote illustrates this problem well:

> In a small town, with a charming market place, citizens respect the local radio station as well as the local watchmaker. One day, the young, inquiring journalist of the newspaper went to the radio station, asking how they knew the exact time. The radio station employee replied that it was because of the watchmaker in the market place. So, the journalist went to the watchmaker with the same question. The answer was: "But that's obvious: from our local radio station!"

I wonder how the journalists from the radio—who I listen to while driving in the morning—would have answered my question: how do they know when to give the message 'it is now 7 o'clock'?

Prior to an in-depth discussion on the traceability of chemical measurements, let us devote a moment to the typical measurements of physical quantities; for example, time. For thousands of years, humans have been dealing with time by preparing calendars, creating patterns of time, and using external machines measuring the passing of time in order to plan each day. The measurement of time is essential both in everyday life and in all fields of science and technology, including chemical measurements. From the very beginning, man has tried to determine the time of day and season of the year, since this was connected with the need to organize life in a society. Initially, these were the observations of the position of the sun and moon, as well as the surrounding nature.

As mentioned earlier, each measurement consists in comparing the unknown quantity to the reference. This must be of the same character as the measured quantity: for example, the length standard must be a section of a certain length; mass standard—a mass of an object; and here is a critical question referring chemical measurements: how we can establish a standard of iron content in human serum?

The creation of an artefact for a standard of time as was done initially (e.g., to measure the length and mass) is not possible. Passing time is, in fact, independent of the will of the individual, and may be passively monitored and in an indirect way—by phenomena that of course is defined as a function of time. A beautiful example was the formerly used so-called 'candle clock,' a burning candle marked with a scale.

In addition, the creation of an artefact of a standard of iron content in blood serum—one that would meet rigorous metrological requirements—is not possible. Thus, a lyophilized blood serum with an established certified content of iron, along with its uncertainty value is used in practice. In this case, the measurement standard is a reference material (RM), accompaning which specification contains information on the traceability of the property (the iron content). Knowing that the chemical measurements are very diverse and include a qualitative and quantitative examination of the vast number of substances, we must realize that providing traceability for each type of measurement is a difficult issue and must always be considered on an individual basis.

4.1.1 The Standard Should Mimic the Measured Quantity to the Best Extent

In practice, in a testing laboratory performing chemical measurements, for calibration of the measuring instrument, the working measuring standards used are most often pure substances; for example, a matrix-free solution containing a known concentration of iron. Matrix reference materials (RMs) (e.g., blood serum containing a known concentration of iron) are used to evaluate a recovery performance of measurement procedure for the determination of total iron in human serum.

Why is the problem of units of measurement and measurement traceability so important? This is mainly due to the essence of the measurement, which is always a process, as already emphasized, in which we compare the unknown quantity with a known quantity. In the case of measuring instruments, we 'teach' the instrument to respond based on the features of the standard of measurement. The process of 'teaching' a measuring instrument is called a calibration. We therefore consider, for example, the calibration of balance or the calibration of spectrometer.

The definition of measurement traceability, in addition to the requirement to ensure an unbroken chain of comparisons, refers to the requirement to attribute uncertainty at all levels of the chain, as exemplified in Fig. 4.2. Hence it is extremely important to know these individual uncertainties in order to incorporate them into the uncertainty budget for a given type of measurement.

The list of commonly used terms in respect to measurement standards and their definition, as given in ISO/IEC Guide 99 is shown in Table 4.2.

All the above-listed standards have their specific place in the metrological hierarchy, each with their associated uncertainty. However, it is worth noting that none of them is either better or worth in a general sense; each plays its specific and important role. The hierarchy of standards reflect the established metrological infrastructure, providing access to the unit of measurement, consistent with the standard of the highest metrological order for all parties (e.g., governments, industry, economy, individual customer). Thus, when buying fruit at the local market, we want the balance used by

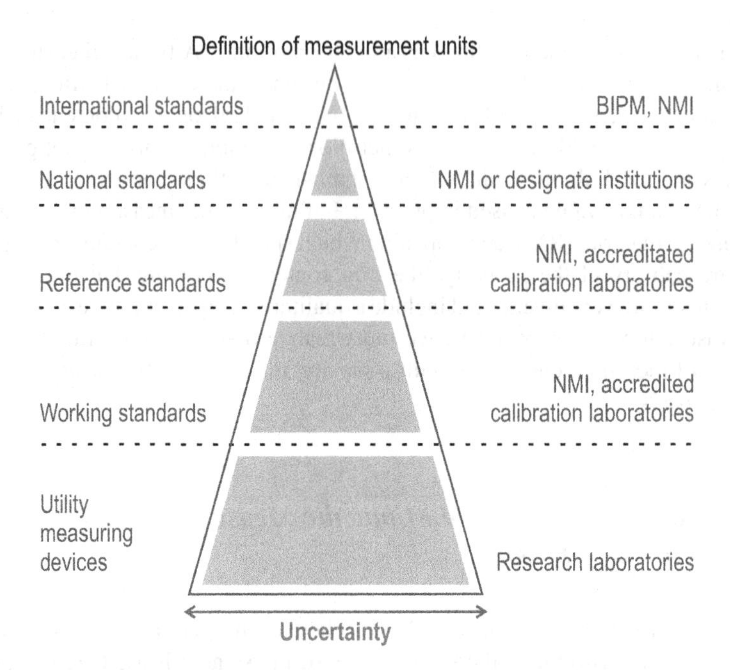

Fig. 4.2 International system of traceability chain with allocated uncertainty of measurements

the vendor to be consistent with international standard of mass unit (the kilogram), and this can be ensured when the balance used in the marketplace was calibrated in the calibration laboratory, which keeps the secondary standard of kilogram, consistent with the standard maintained at the NMI at the given country. Obviously, the uncertainty of weighting, wherein the working weights are used, is greater than the uncertainty when using the weights in the calibration laboratories.

According to the requirements of ISO/IEC 17025:2017, p. 6.4.7, the "laboratory shall establish a calibration programme, which shall be reviewed and adjusted as necessary in order to maintain confidence in the status of calibration." This requirement has several important implications; above all, it requires calibration of all measuring instruments that are used in the measurement process. In the chemical measurements, this certainly includes balance, weight and volumetric glassware. Not required, however, is the calibration of the measuring cylinder, which is used for measuring the estimated volume of liquid volume; or the pipette used to introduce a few drops of the indicator solution before titration.

Whenever a significant effect may be due to temperature (e.g., the temperature of the extraction), it is necessary to calibrate the thermometer used, and whenever the reaction time is essential, it is necessary to calibrate the stopwatch. Thus, it should be highlighted that the calibration applies to those instruments the use of which has a significant impact on the result of measurement. This leaves space for decisions by the chemist, who should determine the significance criteria for the given type of mea-

Table 4.2 List of commonly used terms related to measurement standards

Name	Description
Measurement standard	Realization of the definition of a given quantity, with stated
Primary measurement standard	Established using a primary reference measurement procedure/created as an artifact, chosen by convention
International measurement standard	Recognized by signatories to an international agreement and intended to serve worldwide*
National measurement standard	Recognized by national authority to serve in a state or economy**
Secondary measurement standard	Established through calibration with respect to a primary measurement standard for a quantity of the same kind***
Reference measurement standards	Designated for the calibration of other measurement standard for quantities of a given kind in a given organization or at a given location
Working measurement standard	Used routinely to calibrate or verify measuring instruments or measuring systems
Travelling measurement standard	Standard of special construction, intended for transport between different location
Intrinsic measurement standard	Based on an inherent and reproducible property of a phenomenon or substance
Transfer measurement device	Device used as an intermediary to compare measurement standards
Calibrator	Measurement standard used in calibration

*BIPM: International Bureau of Weights and Measures; WHO: World Health Organization; IAEA: International Atomic Energy Agency
**NMIs: National Metrology Institutions
***Accredited calibration laboratories

surements. Evaluation of how important the influence of the measuring instrument is must be done for each individual measurement. It certainly is not necessary to use a thermometer with a calibration certificate when the measurement procedure within is drying a sample to a constant weight, where the assessment of the effectiveness of drying is carried out on the basis of the weighing results rather than the temperature.

Table 4.3 shows the selected examples for assessing the significance of the effects of the instrument or laboratory equipment for the quality of the result. It should be noted, however, that these criteria must be assessed individually. A good example is ensuring the consistency of measurements; i.e. ensuring the traceability of comparing the measured time with a reference instrument for clocks hanging on the wall in each laboratory room. In most cases, we can say that it would be overzealous to put the effort into ensuring their traceability. Suppose, however, that the analytical procedure requires a sample transfer to the next room, and the documentation

Table 4.3 Selected criteria on the evaluation of the importance of analytical parameters on the final results

Parameters used Step in analytical procedure	Traceability via ...	Influence on the result of measurement
Recovery for CRM, calibration of measuring instrument	Certificate of RMs	Extremely important
Calibration: balance, vessels (flasks, pipets), thermometers	Calibration certificate	Important
Flasks with lower accuracy (e.g., cylinders), pH buffer	Producer specification	Less important
Beaker, Erlenmajer flask, funnel, crucible	Producer specification	Less important

Certified Reference Materials CRM
Reference Material RM

includes recording 'input' and 'output' time for a sample from the room. The time of 'input' and 'output' means the duration of the analytical operations, such as heating, extraction and derivatization. Imagine that the clock in the room II is delayed by 8 min compared with the clock in the room I and III; in this case, the retrace of the conditions of measurement will not be correct.

According to generally accepted practice, the calibration of measuring instruments may be carried out by an accredited calibration laboratory issuing a calibration certificate, which should contain information about the results of measurements with the associated uncertainty. In the case of measurement of physical quantities, the calibration certificate issued by accredited laboratory is sufficient evidence for ensuring the traceability of results.

A more complicated incontinent situation occurs when the measurement of chemical quantities become of interest. This requires usually using various instruments, such as spectrometers and chromatographs with different detectors, especially when we denote qualitative and quantitative composition of complex real samples. For this, refer to the provisions of Clause 6.5.3 of ISO/IEC 17025:2017:

> When metrological traceability to the Si units is not technically possible, the laboratory shall demonstrate metrological traceability to an appropriate reference, e.g.:
>
> (a) certified values of certified reference materials provided by a competent producer;
>
> (b) results of reference measurement procedures, specified methods or consensus standards that are clearly described and accepted by an appropriate authoritative body as providing measurements results fit for their intended use and ensure by suitable comparison.

Thus, participation in a suitable programme of interlaboratory comparisons (ILC) is required where possible.

Those explanations are extremely important for chemists because they refer to the part of the measurement procedure, for which there are no available metrological standards of measurement units. The definition of measurement traceability clearly defines how to assign the appropriate value of the measurement by relating/comparing it to the value of the reference. As has already been emphasized repeatedly, in the case of the measurement of physical quantities, metrological traceability can be ensured through the use of an appropriate measuring device containing in itself information that is consistent with generally accepted unit of measurement. In most cases, the measurement of physical quantities is independent of the type of the object being measured. For example, no matter whether we consider the mass of the sugar, salt or grits, the measurement result depends on the used measuring instrument (that is, the balance) and the used standard (that is, the weights). In this case, the key part of the measuring procedure is the calibration of the measuring device—the balance and weights. One should proceed similarly with the measurement of other physical quantities; for example, the length and temperature.

Directly ensuring measurement traceability in such a way is not possible in chemical measurements. The determination of the content of iron in serum consists of determining the amount of iron present in a specific chemical form in a sample comprising a plurality of other components that may affect the response of the detector. Thus, the measurement procedure comprises the often-complex physicochemical operations; for example, separation of iron ions from the matrix, converting them into a colored complex after the addition of a suitable complexing agent, and then measuring the absorbance of that complex. Obviously, the measuring instrument, in this case, UV–Vis spectrometer, must undergo the calibration. In such a case, the calibration is carried out using the most appropriate standard solutions containing an increasing amount of the analyte (iron compound). In other words, we 'teach' the measuring equipment to respond to the presence of a chemical quantity. In a real sample, beside the substance of interest other chemical species are also present that may affect the behavior of the substance to be determined (e.g., iron ions) during the preparatory step, especially when adding different reagents. A good solution would be to separate the analyte from a matrix of real sample, but here arises another problem: how efficiently are we able to separate it in a quantitative way. All these aspects make the traceability of the chemical measurements much more difficult to achieve. This is namely due to the lack of available reference standards for all possible cases of chemical measurements, which is not the case in the measurement of physical quantities.

If we consider the analytical procedure as a set of successive steps of processing the sample, then for each of these steps we should established the appropriate reference standard, to ensure measurement traceability of the entire procedure. Therefore, the determination and demonstration of traceability in chemical measurements requires consideration of several aspects. How can this be achieved?

First of all, we should:

- Clearly define the purpose of measurements and select the appropriate measurement procedure;
- Describe the measurement procedure in the form of a mathematical equation;
- On the basis of a validation process, demonstrate that all the factors that may affect the final result were accounted for;
- Select proper reference standards for all steps of the measuring procedure;
- Determine the uncertainty that can be attributed to the result of measurement, taking into account the uncertainty of standards and/or calibrations.

In a typical analytical procedure, both measurements are performed—those that can be directly traced to the units of physical quantities, as well as those for which the reference standard can be realized only through metrologically well-characterized chemical substances (both pure and matrix substances). Examples of practices ensuring measurement traceability for the various steps of the analytical procedure are listed in Table 4.4.

Ensuring metrological traceability comprehensibly for chemical quantities is extremely difficult to implement. In practice, the relevant chemical standards of

Table 4.4 Examples of traceability reference for selected activities in analytical procedure

Activity	Instrument/standard	Traceability to …
Weighing	Balance	Unit of mass Certificate of calibration
Calibration of balance	Weights	Unit of mass Certificate of calibration
Dilution	Pipets, volumetric flasks	Volume/unit of mass Certificate of calibration
Measuring of a dose of liquid	Automatic pipets	Volume/unit of mass Producer specification
Measuring of a dose of liquid	Volumetric cylinders	Volume/unit of mass Producer specification
Measuring of a dose of liquid	Syringe	Volume/unit of mass Certificate of calibration or Producer specification
Measuring of temperature	Thermometers	Unit of temperature Certificate of calibration or Producer specification
Measuring of time	Timer, stopwatch	Unit of time Certificate of calibration or Producer specification
Measuring of absorbance	Spectrophotometer	Wavelength Specification of producer
Measuring of pH	pH buffers	Certificate of RM
Sieving	Sieve	The size of particles
Filtration	Filters	The size of particles
Calibration of instruments	Chemical standards	Certificate of RM
Validation of analytical procedure	CRM	Certificate of CRM
Identification of substance	Pure chemical standards Standard spectra or chromatogram	Certificate of substance Library of reference spectra Library of reference chromatograms

Certified Reference Materials CRM
Reference Material RM

pure compounds, or chemical RMs, which enable the value of the property (the content of the chemical substance in a test sample; the identity of the substance) to be transferred between the various institutions and reproduced in different laboratories.

Such conduct does not provide a direct reference to the international system of units (SI) but is consistent with the requirements of ISO/IEC 17025 and policy of accreditation bodies. When the calibration or testing cannot be done strictly in SI units, confidence in the results of measurements is determined by:

- The use of CRMs provided by a competent supplier. The accreditation criteria for Reference Material Producer (RMP) are given in ISO 17034:2016;
- The use of established methods and/or agreed upon references that are clearly described and followed by all parties.

In the measurement of the chemical quantity, the result depends on the applied measurement procedure, and the ability to compare results is often restricted to the same measurement conditions. In addition, the implementation of traceability in the chemical measurements is necessary to correctly define the aim of measurements, taking into account matrix effects, the diversity of the composition of each portion of the sample, the non-homogeneity of the test material and the stability of the sample.

4.2 General Requirements for Metrological Traceability

In order to ensure the comparability of the results given by laboratories worldwide, the common policy should be implemented, which should follow the requirements of ISO/IEC 17025 and ISO 15189. Such a policy is outlined in the ILAC Policy Document P10:01/2013 *ILAC Policy on the Traceability of Measurement Results.*

Metrological traceability requires an unbroken chain of calibrations to stated references, all with assigned uncertainties. Within such a chain, a sequence of measurement standards and calibrations are used to relate a given measurement result to a stated reference,

Important features of measurement traceability include:

- An unbroken chain of comparisons to international or national standards of measurement;
- Documented measurement procedure;
- Documented measurement uncertainty;
- Sound technical infrastructure and technical competence of personnel;
- A reference to SI units, measurement standards or reference measurement procedures containing units of measurement;
- Defined intervals between calibrations.

4.3 Measurement Traceability Over Time

The issue of the constancy of traceability in time, which means maintaining the reference value over the time of storage and of using the standard does not only concern the chemical standards. Many years ago, metrologists observed that the standard kilogram artefact stored in BIPM is changes its mass when compared with the replicates used in the National Metrology Institutes. Although no one could explain exactly why this was happening, the impact of the shifting characteristics of the metrological standard and/or the measuring device must be taken into account in ensuring the consistency of measurements carried out at the given moment. In 1998, in the *Journal of Research of the National Institute of Standards and Technology* published by the American National Institute of Standards and Technology (NIST), an article appeared that focused directly on the issues of standards stability over time [Ch.D. Ehrlich and S.D. Rasberry, Metrological Timelines in Traceability, J. Res. Natl. Inst. Stand. Technol. 103, 93–105 (1998)]. This is especially important if the quantity being measured is not stable with time, meaning that the measuring instrument undergoes drifting or if the reference standard to which the measurement is to be traceable changes significantly with time. In this case, stability refers to both the reference value and the uncertainty of calibration results. Although the authors of this article, Ch.D. Ehrlich S.D. Rasberry, are not chemists, the described phenomenon thoroughly applies to chemical substances—standards of pure chemical substances and RMs (Fig. 4.3).

Over time, during transport and storage, chemical references are exposed to various physical and chemical factors that cause them to change their characteristic properties (Fig. 4.3). Hence it is important to order chemical standards that are distributed

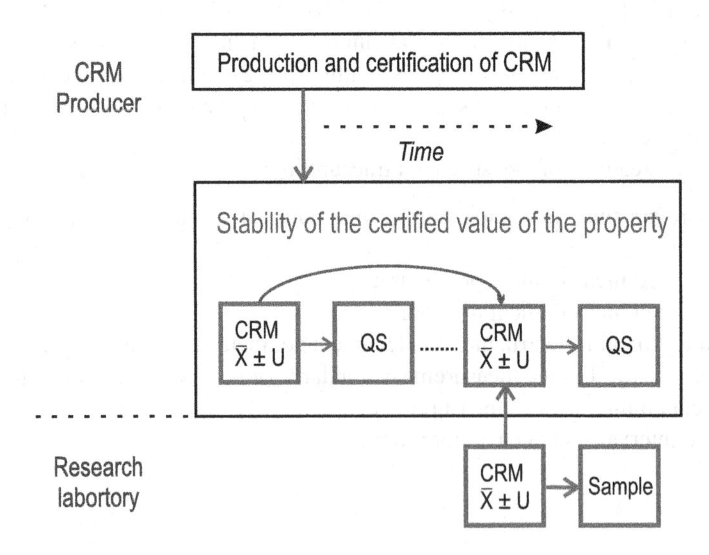

Fig. 4.3 Traceability of chemical standard over time

by competent and reliable producers, acting in agreement with the requirements of ISO 17034. Regarding traceability over time, this means that the producer must periodically determine the value of the certified property, and in case any changes occur must immediately notify its customers or laboratories.

4.4 Compliance with the Requirements of the Reference Material

An ideal RM (standard) is one that is identical in terms of physical properties and chemical composition to the test object. Hence it is extremely important that the producer of an RM specifies as much information as possible in the certificate of material. Among the important information is the homogeneity of the substance. In the case of powder materials, the producer should specify the portion of minimum weighed that provides retaing of certified value. For occlusive matter, used in direct measurements, without carrying out the sample to a solution, the producer should provide information about the homogeneity of distribution of the compounds of interest on the surface. Interestingly, some solids that are considered homogeneous when using a relatively robust measurement technique (e.g., X-Ray Fluorescence; XRF) do not provide the required homogeneity in the application of measurement techniques with higher surface resolution (Laser Ablation Inductively Coupled Plasma Mass Spectrometry; LA-ICP-MS:). This applies, among others, to the standard of various alloys recognized for a long time as homogenous. When using those standards for LA-ICP-MS, the level of homogeneity is no longer sufficient. The situation is similar to the standards of glass, which are commonly used in the examination of the glass composition with the XRF technique. When carrying out measurements using microsampling by laser ablation characterized by a significantly higher surface resolution, these standards do not guarantee the required homogeneity.

4.5 Metrological Traceability in Practice

Assume that the aim is to determine the total iron content in the sample of dried leaves of peppermint. The analytical procedure used in the laboratory comprises the following steps:

- After pre-drying leaves in air conditions and subsequent grinding, weigh approximately 5 g of powder;
- Then dry the sample at a temperature of about 85 °C to a constant weight, so as to remove moisture;
- After drying, weigh accurately on an analytical balance three parallel sub-samples of about 1 g each;

- Transfer quantitatively weighed sub-samples to volumetric flasks (100 mL), pour 20 mL of acetate buffer with an appropriate pH and leave for 20 min, stirring the contents of each flask several times during this time;
- Add the solution of ascorbic acid, to reduce Fe(III) to Fe(II). After completion of the reduction process, add 5 mL of 1,10-phenanthroline and wait another 15 min, so as to allow the formation of a colored complex;
- Fill the flask to the mark.

Before measurements, calibrate the UV–Vis spectrophotometer, using a series of standard solutions containing increasing concentrations of iron, prepared from a stock solution of iron nitrate of known purity. Colorful solutions of the iron complex with 1,10-phenanthroline can be prepared in the same way as the procedure followed in the preparation of sample solutions. Simultaneously, also prepare a blank solution, to which buffer, reducing agent, and a complexing reagent are added. The absorbance of successive standard solutions must be measured. Then, the absorbance of the sample solution of leaves can be measured and the content of iron in solution calculated.

The procedure described above is a relatively straightforward example of laboratory practice. Nevertheless, it allows the most important items related to ensure traceability to be highlighted. In practice, it is worth first distinguishing the physical quantities, for which the establishment of traceability is known. In the above-described example, weighing occurs repeatedly and always allows reference to the traceability to a unit of mass, the kilogram.

The reference to the unit of mass is valid for the following stage of the analytical procedure:

- *Precise weighing of the three portions of about 1 g each.*

In this case, the metrological status of analytical balance and weights is important, and the records in the calibration certificate allow the measurement traceability to the kilogram to be found, as well as the accompanying uncertainty.

WARNING! At the first measurement of sample weight, before the drying, there is no need to show traceability, since this step is indicative of estimated weight (about 5 g), providing the possibility of further weighing of at least 1 g per sample (the procedure requires three sub-samples). Also, the drying conditions (temperature, time) do not need in this case to ensure the traceability of thermometer and clock since the assessment of the correctness of the drying process does not result from the process at a specific temperature (in which case it would be necessary to ensure the consistency of measurement temperature). In this case, the important difference is in mass between successive drying stages.

The use of measuring vessels, flask and pipette

In the case of measuring vessels, it is worth always finding the traceability to the SI unit of mass—the kilogram. Although in practice we use the volume of a flask or pipette, the calibration of measuring vessel is accomplished by weighing the liquid contained in a vessel, taking into account the appropriate correction of temperature for expansion of water and glass. In the case of 'inflow' vessels—for example, volumetric flasks—we weigh the empty container, and subsequently, the vessel is filled to the mark with water. In the case of 'outflow' vessels—for example, graduated pipettes—we weigh the mass of water after pouring it from a pipette. In practice, we weigh the empty vessel, and then the one with a pipetted portion of the water. Measuring vessels can be calibrated independently in the laboratory by weighing water contained in them. In this case, traceability refers to the certificate of calibration of analytical balance and weights. Measuring vessels can be bought with a certificate of calibration, which means that the process has been made by the producer and the user buys a measuring vessel with an appropriate metrological service. Calibrating the measuring vessel is always carried out by weight, which means that regardless of whether we execute this process in the laboratory, use a balance or purchase a service of calibration by a manufacturer of volumetric glassware, the measurement is consistent with the SI unit—the kilogram.

In the case of the described procedure, measurement traceability must be ensured only for the 100 mL volumetric flask, the one in which a sample solution is prepared. Pipettes used for dispensing reagents (buffer solution and 1,10-phenanthroline) do not need the ensured traceability since they are required only for the transfer of the solutions of reagents, which are given in excess.

Figure 4.4 shows the volumetric flask with a volume of 500 mL (nominal capacity) with typical information provided by the manufacturer. According to the requirements of the German Committee for Standardization (DIN; *Deutsches Institut für Normung*) glass measuring vessels should have the following markings:

- Producer (name and trademark)
- Nominal capacity (e.g., 500)
- Tolerance (e.g., ± 0.25)
- The unit of measurement (e.g., mL)
- Calibration information (e.g., 'inflow' calibration)
- Applied temperature (the temperature of the water during calibration)
- Class of the vessels (class A is the highest quality class of measuring vessels).

Preparation of the series of calibration solutions

At this step, iron (II) nitrate of known purity is used, which is weighed and inserted into the flask. To the flask, a portion of the solvent (e.g., 0.1 mol/L HNO_3) is added, which permits the solubility of substances, and then the flask is filled up to the mark.

The process ensuring the traceability for weighing the reference substance is as follows:

- We must ensure traceability of the reference weighing, since it is important to know the exact weight of a reference substance and its molar mass;
- We must ensure traceability of the flask (weight calibration or buying a flask with a calibration certificate from a competent manufacturer).

In order to prepare standard solutions, the iron (II) nitrate with a defined chemical composition and sufficient stability, together with adequate and known purity is used. It is rare that the purity of the reagents should be checked on-site in the laboratory. A common practice is to purchase them from a producer of chemical reagents. In this

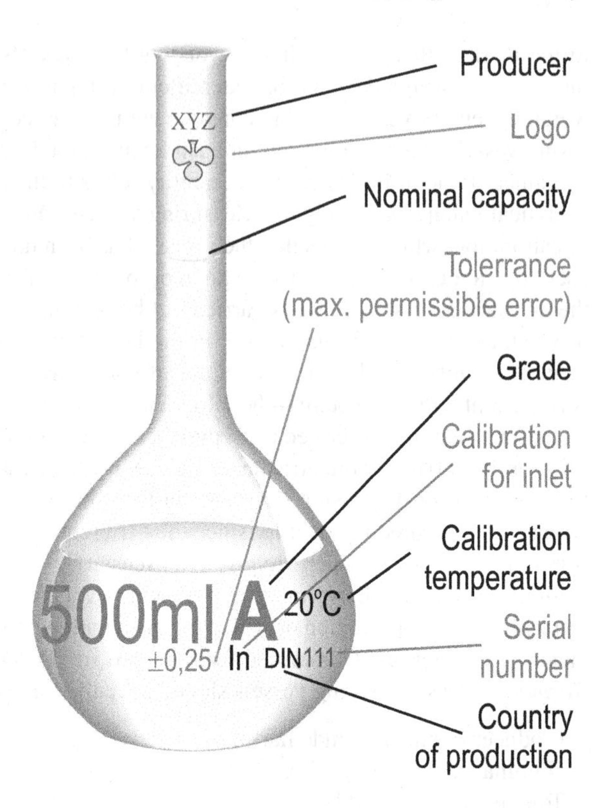

Fig. 4.4 Markings on the volumetric flask

case, we are looking for substance, not only of high but above all, of known purity. Thus, the producer shall provide the certificate of reagent purity and, if possible, a reference to a method for determining the purity, i.e. information about traceability. Knowing the mass of the substance, its molar mass and purity, we can calculate the concentration of iron (II) in a solution prepared in a flask of 100 mL. Knowing the concentration of iron in the following standard solutions, we can prepare the calibration relationship (calibration curve), for which we can clearly demonstrate its traceability.

Standard solutions can also be prepared from the purchased stock solution. The source of traceability can then be a certificate of RM, which should include information on the reference value (the content of the substance), assigned uncertainties and information regarding the certified standard of higher order (higher metrological status).

Determination of total iron content in the leaves of peppermint

We finally approach the most important step associated with the metrological traceability of chemical measurements. As mentioned earlier, in most cases, we do not possess a standard identical to that of the investigated object, which means that we cannot unequivocally compare peppermint leaf samples of unknown iron content to the peppermint leaves standard of known iron content. Perhaps some analysts will argue this because they just saw peppermint leaves in the catalogue of RM, for which a certified iron content was given. If so, they are in a very comfortable but, unfortunately, rare situation.

However, before we content ourselves with buying the available RM on the market, we must ask a few questions regarding the similarity of the material and its characteristics with the test sample, as well as the similarities of the procedure of preparation of samples:

- Are peppermint leaves from plant derived from the same species (trees) as those used by an RMP?
- Is the iron content similar to the content in the test samples?
- Does the reference value relate to the total iron content after complete digestion of the sample or after extraction?

An RM, if it is as far as possible similar to the characteristics of the test sample, may be an adequate source of metrological traceability, provided that the testing is carried out in terms of recovery. These studies, in order to be a carrier of relevant information, should be made according to the same measurement procedure that was used for the samples. The recovery allows an assessment of the impact of the measurement procedure on the final result. It is not important whether the recovery is close to 100% or is more like 68%, as long as the value is repeatable within the range of accepted accuracy and associated uncertainty.

4.6 Summary

Measurement traceability is an essential part of measurements because it allows a result to be related to the recognized measurement standards. The measurement procedures in chemistry are the most complex processes, thus ensuring traceability cannot be considered to be a routine task and must include all elements of the process.

Chapter 5
Calibration

Calibration allows the relationship between the response of detector built-in measurement instrument and the amount/concentration of the measured quantity of any kind to be established.

In the measurement of physical quantities (e.g., mass, volume, time) calibration means assigning a known value of the response to a measuring instrument. The result of a calibration may be formally recorded in a document called a calibration certificate or a calibration report.

Calibration—the operation that, under specified conditions, in the first step establishes a relationship between the quantity values with measurement uncertainties provided by measurement standards and corresponding indications with associated uncertainties and, in a second step, uses this information to establish a relation for obtaining a measurement results from an indication.
Clause 2.39, ISO/IEC Guide 99

Although in some publications, only the first step of the given definition is regarded as a calibration, its broader meaning is now commonly accepted. In the case of chemical measurements, calibration of the measuring instrument covers establishing a calibration relation given in the form of a function (calibration function) or graph (calibration curve). An important issue for calibration is to determine the uncertainty for the individual calibration points (the concentration of the chemical compound) or for entire function (the range of concentration) so that they can be included in the overall budget of the uncertainty of the measurement.

The calibration of balance consists of weighing the calibrating weight, for which the laboratory has a calibration certificate issued by an accredited calibration laboratory. The calibration of volumetric flask consists of weighing the specified volume of liquid; in the case of flasks (vessels of the inflow type), an empty, dry flask should be weighed, then distilled water should be added to the level marked on the neck line

E. Bulska, *Metrology in Chemistry*, Lecture Notes in Chemistry 101,
https://doi.org/10.1007/978-3-319-99206-8_5

and the filled flask should be weighed again. The weight of water in the flask is the difference between those two values. Considering the density of water and including correction for temperature, the volume of given volumetric flask may be calculated.

The calibration of balances or volumetric flasks may be ordered as a metrological service from the producers or calibrating laboratories. In both cases, a document confirming their metrological status is the calibration certificate, in which the values are given for the magnitude and uncertainty of measurement.

Calibration and Measurement Capability (CMC) is considered to be a feature of mutual recognition of national standards.

The best measurement capability (always referring to a particular measured quantity) means the smallest uncertainty that a given calibration laboratory can achieve in practice determining the values of the measured quantity. In the case of measurements of chemical quantities, the calibration correlation is established for specified chemical references. Primary methods of measurement* are considered to be of importance in the chain of traceability as they provide the direct correlation between the abstract definition of a unit of the SI to its practical use in measurement. In the field of chemical measurements, a definition of a primary method of measurement has been developed by the Consultative Committee for Amount of Substance (CCQM), which distinguishes between those methods that measure a quantity directly and those that measure the ratio of two quantities. In several cases, for the methods recognized as primary, where unequivocally described the chemical reaction of known stoichiometry can be used, it is possible to perform calibration by measuring, for example, the mass of precipitate, the volume of the standard solution, the amount of electricity used for the transfer of matter (in coulometry) or the isotopic ratio of stable isotopes.

* The primary method of measurement is one in which the measured quantity is obtained by direct measurement of the base quantity (mass, volume, electrical current, time). Absolute methods include, for example, gravimetry (*weight*), titration (*volume*), coulometry (*electric charge*), isotopic ratio (*current counts*). In the case of trace analysis, absolute methods include neutron activation analysis (NAA) and isotopic dilution mass spectrometry (ID MS). A primary method of measurement is a method for which: (a) a measurement model is completely understood; (b) no empirical factor has to be included in a mathematical equation; and (c) it is possible to realize in a way that achieves a very small combined uncertainty for the given measurand.

In analytical practice, calibration means to determine the relationship between the value of the detector respond relates to the scale on Y (analytical signal) and the value of the content of analyte in a standard relate to the scale on X, generating respective respond. The values of the recorded signals and the known quantity of the given chemical substance in the chemical standard allows for the designation of the calibration relationship in a range of concentrations. In order to draw the calibration curve, the value of the analytical signal is attributed on the vertical axis (Y) and the

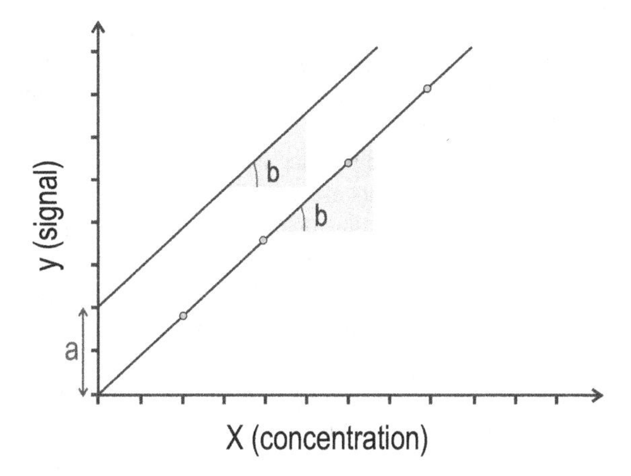

Fig. 5.1 Model graph of calibration relationship (**a** state for the *y*-intercept of the line; **b** state for the slope of the line)

value of the concentration or mass of the analyte on the horizontal axis (X). The most expected case is whenever the relationship between the amount of the analyte and the response of detector is in direct proportion, and a graph of this relationship is linear (Fig. 5.1). In practice, we can describe the calibration graph by several key parameters related to the performance of the analytical procedure: (a) linear range; (b) working range; (c) limit of detection; (d) limit of determination; (e) sensitivity. All of them can vary with the kind of sample (matrix), thus it may need to be evaluated for all types of analyzed objects.

Sensitivity is defined as the slope of the calibration graph and in general, the greater the sensitivity (i.e., the steeper the slope), the more pronounced the difference in concentration.

The calibration dependence is linear usually in a specific concentration range (named the 'dynamic range'), but at higher concentrations, above the so-called upper limit, it is often curved. The calibration relationship does not have to be ideally linear and this can be evaluated by statistical means; for example, fitting by higher order mathematical model. The working range, usually greater than linear range, covers the concentration range where the results can be given with acceptable uncertainty.

The limit of detection (LoD) and the limit of quantification (LoQ) are defined as the minimum concentration of analyte that can be detected with statistical confidence and as the lowest concentration of analyte that can be determined with an acceptable uncertainty, respectively. The value of LoD can be estimated using the solution of blank or sample containing a very small concentration of analyte, since the value of LoQ should be determined by using appropriate chemical standard. An example of model graphs of the purely linear calibration curve is shown in Fig. 5.1.

This graph can be described by the calibration function, $\boldsymbol{y} = \boldsymbol{b} \times \boldsymbol{X} + \boldsymbol{a}$, where: $\boldsymbol{y}$ is the value of the signal, $\boldsymbol{X}$ is the concentration, $\boldsymbol{b}$ is the slope, and $\boldsymbol{a}$ is the coordinate point of intersection of the line with the ordinate axis.

The analyte concentration in a test sample is determined by interpolation of the signal obtained for the sample to the Y-axis corresponding to the value of an analyte on the X axis [the second step in the definition of calibration (*Clause 2.39, ISO/IEC Guide 99*)].

The calibration dependence is often established with a number of standards, and the measurement points are used for plotting the respective line. In practice, points are not positioned in the ideal straight line, a certain spread of data results from random errors. Therefore, a regression line is carried out through the designated measurement points—the line of best fit to the data points. The degree of coherence of measurement points to a straight line is determined by the correlation coefficient. In principle, the graph of the calibration dependence should start at the origin, i.e., in the absence of an analyte in a sample, the measuring point should correspond to the coordinates $\{0.0\}$. Considering, however, that the measurement result is affected by the noise of the measuring system and the fact that a sample blank can contain trace amounts of substance (non-removable), frequently the intersection with the Y-axis is not at zero, but slightly above zero (a factor in the equation depending on the calibration).

In the case of the linear dependence between C and y, two points are sufficient to establish the graph, but in analytical practice, it is recommended to use three or even five standards with increasing concentrations of an analyte. A larger number of measurement points results in reducing the impact of random errors. In addition, it allows a better evaluation of the dependencies. If the relationship is nonlinear, using a larger number of measurement points enables two or more linear ranges over the entire range of concentrations to be distinguished.

The correlation coefficient is used to assess the degree of linearity of the dependence of two variables. If the calibration is judged according to the linearity of the analytical signal (the response of the instrument) to the concentration or mass of the analyte (the standard quantity), the correlation coefficient r for the variables x and y is determined by the following equation:

$$r = \frac{\sum_{i=1}^{n}[(x_i - \bar{x})(y_i - \bar{y})]}{\left\{\left[\sum_{i=1}^{n}(x_i - \bar{x})^2\right]\left[\sum_{i=1}^{n}(y_i - \bar{y})^2\right]\right\}^{1/2}} \quad (5.1)$$

where: $x_1, x_2, \ldots, x_n$, and $y_1, y_2, \ldots y_n$ represent the coordinates of the points, x and y, and are the mean values of x and y, and Σ means the sum of the respective elements. The correlation coefficient factor ranges from -1 to $+1$. The value of $|1|$ indicates a perfect correlation (which is possible only in the case of a straight line represented by two points), and the value 0 indicates no correlation at all. The values of correlation coefficient r can be positive or negative depending on the slope of the calibration relationship. In practice, in chemical measurements, the calibration correlation is characterized by a positive slope and the correlation coefficient r has a value above 0.9899 (usually, the value is given to an accuracy of four decimal places).

Linear regression is a mathematical method that allows the course of a straight line of best fit to the data points to be determined, for which the coefficient of linear regression showed a satisfactory degree of linearity. Linear regression allows the calculation of the coefficients b (slope) and a (the intercept point of the ordinate axis) for the straight line of best fit to the data points. The most commonly used mathematical algorithm is a least squares method, where a and b are selected such that for the equation $y=b \ldots x+a$ the smallest value was the expression $\Sigma\ (y_i - y)^2 = \Sigma\ (y_i - a - b \times x_i)^2$, where y_i and x_i indicate values of consecutive measurement of the analytical signal (y_i) and the concentration of analyte in the standard (x_i), respectively.

$$a = \frac{\sum_{i=1}^{n} y_i - b \sum_{i=1}^{n} x_i}{n} \tag{5.2}$$

$$b = \frac{n \sum_{i=1}^{n} x_i y_i - \sum_{i=1}^{n} x_i \sum_{i=1}^{n} y_i}{n \sum_{i=1}^{n} x_i^2 - \left(\sum_{i=1}^{n} x_i\right)^2} \tag{5.3}$$

Establishing a line graph with the least squares method requires meeting the relevant criteria, including the equality of all measuring points used to determine this relationship. This means that each point of the graph should be the average of several measurements, and the standard deviation should be the same for all points. Another condition is that the uncertainty of measurement of weight or determination of the concentration of the analyte was small enough to be omissible. Besides that, a normal (Gaussian) distribution of results is assumed. In practice, it is not possible to check whether the measurement points meet the above criteria, but generally, it is assumed that they do.

If the relationship between the analytical signal and the mass or concentration of the analyte is not linear, it is possible to use more sophisticated calculation methods, i.e., curvilinear or logarithmic regression.

5.1 Bracketing Over a Range of Concentrations

In routine measurements, where there is a need for assays for numerous samples and the expected range of analyte concentrations is known or can be estimated, the use of the bracketing method for a specified range is very useful. In this case, the calibration relationship is determined for the concentration range using a standard at a concentration slightly smaller than the smallest expected content of the analyte and slightly greater than the expected highest content of the analyte. Assuming a linear relationship between both measuring points for the two standards, the content of the analyte in the sample is determined by linear interpolation.

5.2 Evaluation of Recovery with Reference Materials

In the case of chemical measurements, pure chemical substances are often used to calibrate the measurement instrument. They can be used as a solid substance or as a solution, the latter is commonly named the 'standard solution.' Thus, the metrological traceability in chemical measurements can be achieved by linking the respond of instrument to known amount of pure chemical standards (solids or dissolve, depending on the instrumental technique used for performing measurements), knowing its atomic and molecular mass in moles, the dimensionless SI unit. In practice, we can see two main limitations of such an approach: first, it is not possible to acquire completely pure chemicals; second, it is not possible to isolate the analyte completely from the samples, meaning to separate the analyte from the matrix. In addition, setting the calibration relationship for pure chemical substances (the first step described in the definition of calibration given in VIM) and then determining it according to the content of an analyte in a real analytical sample (the second step described in the definition of calibration given in VIM), in a case when measurement can be interfered with by matrix component can lead to systematic errors. In this case, it is necessary to check the influence of the matrix of the sample on the analytical signal, since it may alter its physical and chemical properties, or cause undesired interference and, consequently, induce a change of values a and b of the calibration function. The influence of the matrix can be archived via evaluation of the recovery.

Ideally, the best approach is to correlate the calibration relationship for references which should be identical in chemical composition and physical form to the test objects (samples). Establishing the calibration dependence with the use reference material of identical composition to that of real test samples is in most cases difficult to implement, due to the lack of the wide range of them. Thus, it is recommended to use reference materials as similar as possible to test samples in respect of the content of analyte and the matrix composition. The performance of the entire analytical procedure is then evaluated by determining the concentration of analyte of interest from the calibration and comparing to the value specified in the certificate completing by calculation of the recovery. Compliance of the results within the accompanying uncertainty is a confirmation of the absence of significant interference. If values differ statistically from the certified value, the appropriate budget of uncertainty should be taken into account or the corresponding recovery factor should be included into the model equation.

5.3 Method of Standard Addition

In cases where it is not possible to apply standards so as to match, as closely as possible, the sample matrix, and if it could be expected that the components of the sample significantly affect the detector response (causing interference), the determination of an analyte based on the relationship of the calibration set for the references of pure

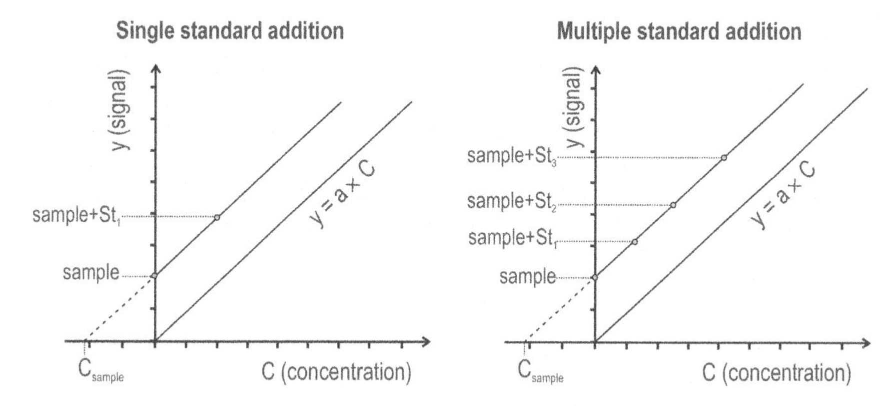

Fig. 5.2 Graphical example of the method of standard addition

substances can result in underestimation or overestimation of the result. Therefore, it should be consider the overcome this problem by adding standard to the sample as to calibrate in the presence of the sample matrix. The so-called *method of standard addition* is very useful, especially when the composition of the sample is not easy to mimic or not known and/or matrix components cause significant interference. In such cases, it may be helpful to use the calibration method of standard additions. In practice, a single or multiple addition of a standard can be applied (Fig. 5.2).

The addition of pure chemical standard of the same kind as the analyte can be called 'spiking.'

It sometimes seems to be useful to follow the calibration of the measuring instrument with the use of a set of standards of a pure chemical substance by a standard addition method. This enables comparison of the slope of both graphs and evaluation of whether in fact the interference effects may appear in the test samples. If the slope of both graphs are the same within the accepted uncertainty, it can be assumed that no interference will disrupt the results for the test samples.

Single standard addition method

The method of single standard addition involves measuring the analytical signal for the test sample with unknown analyte content, and then measuring the analytical signal for the sample to which a known amount of analyte was added. The amount added to the sample analyte (standard addition) should be close to its expected content in the sample. Thus, two measurements are undertaken for the given test sample, before the addition of the standard and after the addition of standard. This procedure can be used when the chemical standard added is identical to the analyte originally present in the test sample. Note that the dilution effects should be assumed to be significant when the volume of sample solution, as a result of the standard addition, increases by more than 1%. Moreover, the method of standard addition can be used only when a linear relationship exists between analytical signal y and the concentration of analyte C.

In the method of single standard addition, the determination of an analyte is usually carried out by the graphical method (Fig. 5.2). Two measured signals (sample before and after addition of the standard) are used for plotting the graph of respond against concentration of analyte added. Then the negative intercept on the x-axis, at $y = 0$, represents the concentration of the analyte in the sample solution.

The method of standard addition is a useful approach when external calibration with the set of standards of pure chemicals is not possible because the response is affected by the sample matrix.

Multiple standard addition method

In some cases, improved accuracy of the result can be achieved using the method of multiple standard additions. In this case, we make a number of additions of standard (more than one) to the test sample solution containing the analyte, and then measure the resultant increase in the response of detector after each addition. The results can best be presented graphically. By extrapolating the straight line back to $y = 0$, we can obtain the measure of the value of concentration of analyte in the test sample before any addition.

5.4 The Advantages of the Standard Addition Method

- Using the standard addition method is advantageous when there is no information about the matrix, which could cause interferences;
- The standard addition method can be recommended for an analytical task with many samples of different composition of the matrix. It would then be uneconomical to prepare many calibration graphs that could mimic the composition of the different samples.

5.5 Limitations of the Standard Addition Method

- The standard addition method can be used only if the analytical signal is directly proportional to the concentration C or its function, for example, $\log C$. The calibration relationship is not always linear, so extrapolation may result in overestimation or underestimation of the results.

- The chemical form or the behaviour of the added analyte are not always identical to that of the analyte present originally in the test sample.
- The standard addition method does not eliminate the errors resulting from the presence of additive interferences, causing the parallel shift of the analytical graph as well as those caused by human error.

5.6 The Method of the Internal Standard

In analytical practice, results depend largely on the measurement's parameters as well as on the composition of the sample. For the calibration approaches described above, the assumption was made that the quantity (mass or volume) of standard or test sample was accurately known and the parameters that affect the overall measurement procedure were constant. There are, hoverer several analytical procedures where those assumption are not valid, it is not possible to ensure the quantity of the sample taken for measurements or it is not possible to guarantee the stability of all parameters involves.

In such cases, the methodology that could overcome these problems is the *method of internal standard*. In this method, the response of the detector (analytical signal) given by the analyte of interest is compared with that given by another element or compound of known concentration (named the 'internal standard'), which is present in the sample. Although the internal standard may already be present in the sample in its original constitution, it is often added to the sample before performing measurements.

> The method of internal standard offers very high accuracy and precision, mainly due to the fact that the analytical signal for analyte and standard are both measured in the same portion of the sample and at the same time.

In fact, the method of internal standard involves tracking the signal of the internal standard, simultaneously with the signal of the analyte, which allows for compensation of random errors of the measuring system and/or its variation over time and changes in the composition and/or physical properties of the sample.

As an internal standard, a substance is chosen that does not react chemically with the sample (in case it is added to the sample), nor interfere in any way with the analyte. In contrast to the standard addition method, knowledge of the content of the substance acting as the internal standard is not essential. During the measurements, a monitored signal is obtained for the internal standard, and assuming its concentration is constant, it is believed that any change in the value of the signal indicates a change in the measurement conditions. Another assumption is that the change in the conditions affects to the same extent the signal of the substance acting as an internal standard, and the signal from the presence of the analyte. The calibration dependence, in this case, is a graph wherein on the Y-axis the ratio of the value of the analyte signal to internal standard signal is mapped and on the X-axis the mass or concentration of the analyte is mapped. As previously mentioned, an internal standard may be added to the sample or may already be present in the sample.

Assuming the influence of various instrumental and experimental parameters on the analytical signal obtained, the ideal calibration should meet the following conditions:

- An identical chemical environment of the analyte in the samples used for the calibration and in the sample;
- The same procedure for the preparation of calibration solution and the sample;
- Identical measurement conditions for calibration solutions and samples.

Thus, the method of the internal standard approximate to the best extent the calibration conditions to the above-mentioned requirements.

In the method of the internal standard, the analytical signal—defined as the ratio of the signal obtained for an analyte that of internal standard—eliminates the influences that affect in the same or similar manner the signals of both substances. The principle of the internal standard method is that a fixed amount of internal standard is added to all solutions/samples and the analytical signal is proportional to the analyte in the sample. The internal standard method allows the influence resulting from changes in concentration of the analyte at the time of preparation procedure (incomplete extraction, change of the volume) to be reduced; eliminates the influence of some interferences related to the composition of the sample (e.g., spectral interference); and reduces the influence of measurement conditions on the analytical signal (measurement of the signal for internal standard and the analyte in the same measurement cycle). It is worth noting that the internal standard method is very useful in measuring techniques, in which it is possible to obtain signals for more than one component of the samples in one measurement cycle.

- Internal standard should exhibit similar physicochemical properties as the analyte;
- Internal standard should not interact with both the analyte and other component of the sample.

5.7 Internal Standardization

In some analytical cases, knowledge of the relative contents of the sample components is important. If so, the content of all detected components are normalized to 100% or to 1; thus the content of the individual component is expressed as a percentage or fraction. The method of internal standardization is used in those cases where

Table 5.1 Total uncertainty resulting from the calibration relationship depends on several factors

Source of uncertainty	Comments
Uncertainty of the calibration relationship	Decreases with increased number of calibration points
Uncertainty of measurement result for test sample	Decreases with increased number of repetitions
Sensitivity of the method	Uncertainty decreases with increased sensitivity (higher value of factor b)
Uncertainty of the b in calibration equation	Decreases with increased concentration range Note: the uncertainty is the smallest for the measurements carried out in the central part of the linear range

the relative proportions of the components are more important than the absolute contents of the individuals.

It should be noted, however, that the use of internal standardization runs the risk of making mistakes when we do not take into account the content of certain ingredients (e.g., the XRF measurements does not take into account the content of trace elements), or when in the chromatographic methods we assign the content of different components of a sample from the peak area corresponding to that individual in relation to the entire area of all peaks (not taking into account that the sensitivity of the detector response could be different for each sample component).

5.8 The Quality of Calibration of Analytical Procedure

The quality of the calibration relationship determination depends on:

- The precision of measurements used for calibration;
- The metrological quality of used chemical standards;
- The accuracy of the evaluation of the analytical signal of the test sample delivered from the calibration relationship.

In practice, the calibration relationship should be presented so that it is possible to read the measurement results for the sample with sufficient accuracy and uncertainty. In order to take include this in the uncertainty budget of the entire analytical procedure, it is important to take into account the uncertainty of the reference values and the dispersion of measurement results obtained for each standard (Table 5.1).

The calibration relationship slope can fluctuate over the time of measurement (Fig. 5.3). For this reason, it is essential to establish a process of systematic monitoring of the parameters of the calibration curve. In practice, it is convenient to assess the slope with the use of selected standards, usually at a middle concentration corresponding to the central part of the straight (dynamic) range of calibration dependence.

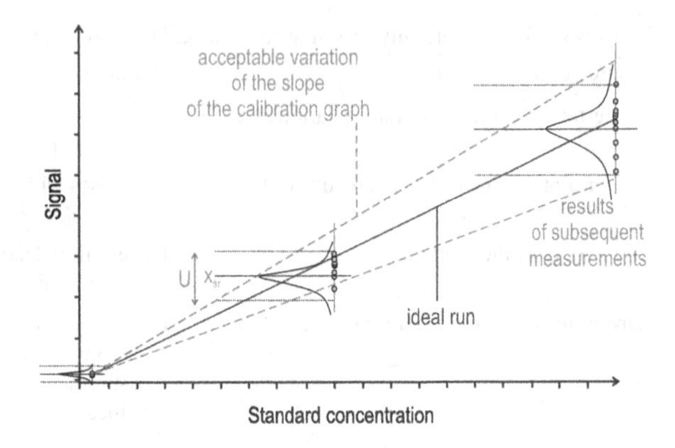

Fig. 5.3 Acceptable variation of the slope of the calibration graph

When conducting routine measurements, it is recommended to control the stability of the slope, at least before and after each series of measurements, and at a large number of samples during a measurement; for example, after each 20th sample. Control may be performed in a simplified manner by means of one or two standards. It is necessary to verify the graph in the whole range of concentrations whenever changing the measuring equipment and conditions for conducting measurements.

Chapter 6
Reference Materials in Chemical Measurements

Certified reference materials (CRMs) are essential in the measurements of chemical properties. They reflect the content of the analyte in the matrix while retaining all specific interactions in the examined chemical system.

In previous chapters, the general issue of metrological traceability as well as calibration has been discussed. The purpose of this chapter is to focus on the chemical standards, namely the reference material (RM) used for achieving traceability in chemical measurements. The discussion presented in this chapter on chemical RMs and their use in providing traceability relates primarily to testing and calibration laboratories that meet the requirements of ISO/IEC 17025:2017.

The main feature of the measurement result is its traceability to the relevant standard (the reference). As previously stressed, in the case of chemical measurements, providing measurement traceability is much trickier compared with measurements of physical quantities. As repeatedly stated, this is due to the fact that in the chemical measurements, the result depends not only on the calibration of the measuring instrument but also on the nature of the test object and the manner in which the object is prepared for the measurements. In the previous chapter, methods for how to establish the appropriate references (metrological traceability) were presented for the subsequent quantities affecting the result. In this context, for some steps of the measuring procedure it is possible to trace back to SI units, primarily through the purchase of metrological services in the relevant accredited calibration laboratories (e.g., balance calibration, purchase of a class A volumetric flask). A direct reference to the SI is not possible for the result obtained using the measurement procedure in which the test object (sample) is subjected to various physical and chemical operations. Therefore, an important part of the activity of chemical laboratories is the use of chemical RMs.

RMs, for which a certified value of the property is known, are a tool to ensure the traceability of chemical measurements. Assuming that for each chemical task a suitable certified reference material was available, it would be possible to establish an infrastructure similar to that existing for the measurement of physical quantities. Unfortunately, both the measurement procedures used and the variety of objects are

E. Bulska, *Metrology in Chemistry*, Lecture Notes in Chemistry 101,
https://doi.org/10.1007/978-3-319-99206-8_6

much more diversed. In this case, the specific techniques used for measurement, the complexity and diversity of test samples and often an extremely low content of the analyte all add to the complexity of the situation, as well as necessitate the use of a multi-step process of sample preparation.

Reference materials (RM) and certified reference materials (CRM) are an extremely important tool for obtaining reliable results of measurements of the chemical quantities.

6.1 Basic Definitions and Requirements

Reference material (RM) is a generic term to describe a group of materials used as transfer standards in chemical measurements. Transfer standards are those that carry metrological information (e.g., identity or content of substance) and can be used for the calibration of measuring instruments. RMs of various kind are used for establishing the traceability via calibration, validation of measurement procedures, optimization of analytical as well as being used for different levels of quality control results (e.g., control charts, interlaboratory comparisons). Despite their wide range of possible application, they should only be used for a single purpose in a given measurements.

The general requirement for establishing the traceability via calibration is to ensure that they are traceable to the SI system of units. However, in the case that certain calibration cannot be strictly made to SI units, traceability could be established to CRMs provided by a competent supplier in order to give a reliable physical or chemical characterization of a material or to specified method and consensus standards that are clearly described and agreed upon by all parties concerned (as described in Clause 6.5.3 ISO/IEC 17025:2017).

The general requirements for the reference material producers (RMP) are listed in ISO 17034:2016, where a number of criteria for reference material (determined uniformity, stability, the stability of certain properties, metrologically properly characterized reference value and its uncertainty) are described. In case those criteria are fulfilled, RM may be accompanied by the certificate of RM. The necessary information, which must be indicated on the certificate of a CRM, are: the name and unique identification of RM, the intended use, instructions for handling, period of validity and storage conditions, property value and associated uncertainty and metrological traceability of the certified values (as the proof of traceability).

Reference material (RM): materials sufficiently homogeneous and stable with reference to specified properties, which has been established to be fit for its intended use in measurements or in examination of nominal properties.
Clause 5.13, ISO/IEC Guide 99

Several notes are added to the definition of RMs, given in VIM 3, which specified various requirements. It is highlighted that a given reference material can only be used for either calibration or quality assurance in a given measurement. The RMs with assigned quantity values can be used for calibration or for establishing a measurement trueness; those without assigned quality values can be used for establishing measurement precision.

There are various kinds of RM; for example, pure chemical substances with specified purity, standard solutions of pure chemicals with specified concentration of given substance, matrix RMs with analyte originally present and with a specified concentration, matrix RMs with added (spiked) known amount of analyte.

Certified reference materials (CRM): reference material, accompanied by documentation issued by an authoritative body and providing one or more specific property* values with associated uncertainty and traceability, using validated procedures.
Clause 5.14, ISO/IEC Guide 99

*properties can be qualitative (e.g., identity of substance) or quantitative

CRMs and RMs are widely used in chemical laboratories. If used properly, they can provide important information about the quality of the results. CRMs in the form of standards of pure substances are most commonly used for the calibration of the measuring instruments, thus ensuring metrological traceability. CRMs, especially those mimicking the matrix of test samples, are used for validation—for the evaluation of the recovery of the applied analytical procedure and to determine the accuracy of the measurement results.

Apart from the definition of CRMs given in VIM 3, several reference material producers (RMPs) use their own trade name to described their RMs. For example, ERM® is the registered trademark for **European Reference Materials**. The ERM® concept is a joint collaboration of major European RMPs who guarantee to: apply the principles currently available described in ISO 17034 and ISO Guide 35 for the production of CRMs; demonstrate rigorously homogeneity and stability for all materials and guarantee the certified value for every single unit over the complete shelf life of the materials; be transparent in their approach to the production of CRMs. Another example is the NIST **Standard Reference Material**® (SRM)—a CRM issued by NIST that meets additional NIST-specific certification criteria and is issued with a certificate or certificate of analysis that reports the results of its characterizations and provides information regarding the appropriate use(s) of the material. Note: an SRM is prepared and used for three main purposes: (1) to help develop accurate methods of analysis; (2) to calibrate measurement systems used to facilitate the exchange of goods, institute quality control, determine performance characteristics, or measure a property at the state-of-the-art limit; and (3) to ensure the long-term adequacy and integrity of measurement quality assurance programs. Another approach of NIST is link to the production of **NIST Traceable Reference Material**^{TM} (NTRM), an RM with a well-defined traceability linkage to existing NIST standards for chemical measurements. This traceability linkage is established via criteria and protocols defined by NIST to meet the needs of the metrological community to be served. RMPs adhering to these requirements are allowed use of the NTRM trademark, which may be recognized by a regulatory authority as being equivalent to a CRM.

An important international forum dealing with RMs and CRMs is the Committee for Reference Materials REMCO, acting within the framework of the International Organization for Standardization ISO. Committee ISO/REMCO was founded in 1975 and presently ISO/REMCO is composed of 32 participating and 38 observer members. The main aims of the activity are related to establishing concepts, terms and definitions related to RMs; to specify the basic characteristics of RMs as required by their intended use; to propose actions on RMs required to support other ISO activities; to prepare guidelines for ISO technical committees when dealing with RM issues. ISO/REMCO cooperates with many institutions, including regional and national standardization organizations and metrological institutes. The effect of the activities of ISO/REMCO is evidenced by a number of guides and technical reports issued in recent years, outlined below.

ISO Guide 30:2015 *Reference Materials. Selected Terms and Definitions,* provides definitions of the following terms: reference material, certified reference material, candidate reference material, matrix reference material, sample and minimum sample size, as well as production batch (lot), characterization and value assignment, homogeneity, stability, lifetime and period of validity and many more items related to RMs.

ISO Guide 31:2015 *Reference Materials. Contents of Certificates, Labels and Accompanying Documentation,* aims to guide RMPs in preparing clear and concise documentation to accompany their RM. This information can be used by RMs users and other stakeholders in confirming the suitability of an RMs or CRMs. This guide also contains the minimum requirements for a label attached to the RM container.

ISO Guide 33:2015 *Reference Materials. Good Practice in Using Reference Materials,* describes good practice in using RMs and CRMs by laboratories in measurement processes. These uses include the assessment of precision and trueness of measurement methods, quality control, assigning values to materials, calibration, and the establishment of conventional scales. It provides important information on the characteristics of various types of RMs in respect of their different applications.

ISO Guide 35:2017 *Reference Materials. Guidance for Characterization and Assessment of Homogeneity and Stability,'* which explains concepts and provides approaches to the following aspects of the production of reference materials: the assessment of homogeneity; the assessment of stability and the management of the risks associated with the possible stability issues related to the properties of interest; the characterization and value assignment of properties of a reference material; the evaluation of uncertainty for certified values; and the establishment of the metrological traceability of certified property values.

ISO/TR 79:2015 *Reference Materials. Examples of Reference Materials For Qualitative Properties* summarizes the state of the art of the production and certification or characterization of qualitative property RMs. The investigation of nominal properties is referred to differently in various specialized areas (examination, classification, identification, testing, observation, etc.).

ISO Guide 80:2014 *Guidance For the In-House Preparation of Quality Control Materials (QCMs)*, outlines the essential characteristics of RMs for quality control (QC) purposes, and describes the processes by which they can be prepared by competent staff within the facility in which they will be used (i.e. where instability due to transportation conditions is avoided). The content of this guide also applies to inherently stable materials, which can be transported to other locations without risk of any significant change in the property values of interest. Those involved in QCM preparation should have some knowledge of the type of material to be prepared and be aware of any potential problems due to matrix effects, contamination, and so on.

ISO/TR 10989:2009 *Reference Materials. Guidance On, and Keywords Used For, Reference Material Categorization,* covers: the results of a study into, and comparison between, existing classification and categorization schemes for RMs, the development of RM features and characteristics upon which a harmonized and consistent categorization scheme could be based, and approaches for making the categorization scheme adaptive to new RM needs and developments.

ISO/TR 11773:2013 *Global Distribution of Reference Materials*, contains an inventory of problems and recommendations related to the transport, import and export of non-nuclear, non-radioactive RMs, specifically for the packaging, labelling, and documenting of the shipments in order to comply with legal requirements.

ISO/TR 16476:2016 *Reference Materials. Establishing and Expressing Traceability of Quantity Values Assigned to Reference Materials*, discusses and specifies the general principles of establishing the traceability of measurement results laid down in the Joint BIPM, OIML, ILAC and ISO Declaration on Metrological Traceability, in particular for values assigned to CRMs. The document covers a study into the existing principles and requirements for the traceability of the value assigned to the property of a (C)RM, the development of a sensible, widely applicable approach to the understanding of the traceability of a value assigned to (C)RM property, as well as recommendations on how traceability should be established, demonstrated and reported on certificates and other documents accompanying (C)RM.

Detailed requirements that are the used for the accreditation of RMPs are described in the document ILAC-G12:200. 'Guidelines for the requirements for the competence of reference material producer.'

6.2 Production of Reference Materials

In between experts dealing with RMs, discussions are also conducted on the forthcoming document that is to include guidelines for manufacturers and users of the RMs located on the lowest level of the metrological hierarchy, such that they can be used in the laboratory in routine measurements of quality control. In practice, various terms, such as quality control materials' or 'laboratory reference materials' are used. But regardless of the name used, it is important to establish widely accepted requirements for materials with the lowest metrological hierarchy, fulfilling at the same time an important role in the daily quality control to the economically justified level.

It is worth noting that while the number of documents related to RMs is not large, they include very different areas: from production to their use in laboratory practice. For the analyst, of course, the most important are the requirements for the proper use of RMs. However, one should also be aware of what the requirements of manufacturers of these materials are, where the results of these requirements are derived from and what the impact is of the manufacturing process on the metrological value of the product. From the user's perspective, an important requirement with regard to the chemical standards and RMs are the competences of the manufacturer. RMs should assure have, where possible, a traceability link to the units of measurement SI or CRMs. It is worth remembering that the production of chemical RMs is a complex process, involving primarily the proper preparation of the material to demonstrate its homogeneity, stability, and the characteristics of accuracy and measurement traceability of the reference value. It is also worth noting the problems related to the acquisition of a suitable material; for example, the contents of medicinal substances in plants vary significantly depending on the location and time of harvest; and animal tissues, for example, are difficult to obtain.

The production of matrix RMs is a highly specialized activity. The production of a single type of material requires huge investment, and the income from the sale of the material rarely fully covers the cost of its production, certification, and storage. The process of planning the processing of the material, its production, certification, and how to determine the conditions of transport of the material to the customers and how it is stored is extremely important. Problems related to the production and certification of RMs are not always known to their users. But it is worth knowing, at least in outline, what difficulties the producers must cope with, since this will allow for a better understanding of the high costs and the proper use of materials in chemical measurements. The manufacturing process involves many steps: acquisition of a substantial portion, often several hundred kilograms of the material; crushing; grinding; freeze-drying; and sterilization. The manufacturer of the material should guarantee its stability within the validity period, and this requires systematic monitoring of the selected property at the time of selling. The high cost of production of RM, however, is fully justified. The production itself requires preparation of appropriate production lines and modern equipment (sometimes it is necessary to prepare equipment designed specifically and exclusively to produce the type of RM in question) and the employment of highly qualified personnel. After the production, the RM must be stored under appropriate conditions, and the manufacturer is also obliged to periodically check whether the reference value has not changed. A single batch of produced material is often sold over many years, which significantly increases the cost of the individual package and extends the amortization periods for investment.

It is also important to know how different RMs are certified by individual manufacturers. For CRMs, the reference value can be determined in different ways, for example:

- Assigning a value a priori (by formulation);
- Assigning a value derived as a result of interlaboratory comparisons;
- Using primary or reference measurement procedure.

In any case, the reference value should be accompanied with a corresponding uncertainty. It is also important to consider how close the certified value meets the requirements described in the definition of measurement traceability. Where it is possible to set the value with the primary method, the measurement traceability is ensured with the clearly described measurement procedure. In other cases, measurement traceability is related to the conventionally accepted values for which it is not always possible to refer to the standard of a higher order.

6.3 Types of Reference Materials

RMs can be divided according to their position in the metrological hierarchy. It should be noted that the position of the given material within the level in this hierarchy does not relate directly to their usefulness and quality. The hierarchy of metrological patterns is considered for the allocation of their reference value and assigned uncertainty. The metrological quality of the standard/reference material must always be selected for a given purpose and justified both logically and economically.

Primal measuring standard is a standard that has the highest metrological quality and whose value is accepted without reference to other higher standards of a given quantity. By contrast, the secondary measuring standard is one whose value is determined by comparison with the primal standard of a given quantity.

In the measurements of physical quantities, the most common primary standard is associated with the respective natural phenomena. For example, the primary standard for time is the second defined by fixed numerical value of the cesium frequency $\Delta\nu_{Cs}$, the unperturbed ground-state hyperfine transition frequency of the cesium 133 atom, as 9,192,631,770 when expressed in the unit Hz, which is equal to s^{-1}. The realization of this unit is carried out by the respective laboratory at BIPM and selected NMIs.

In the measurement of the chemical quantity, primary RMs, the highest in the metrological hierarchy, are materials whose properties are determined by definitive methods (also called primary methods). Primary RMs produced by NMIs are traced to SI units and are assigned an uncertainty in accordance with the procedure described in the respective ISO GUM Guides. Additionally, they are otherwise subject to control within the framework of interlaboratory comparisons (i.e. the key comparisons).

Lower in the metrological hierarchy are secondary RMs, which are prepared by accredited calibration laboratories or accredited RMPs. On the lower levels of the metrological hierarchy are materials prepared in the laboratory and used in the daily processes of quality control.

Commonly used terms for the RMs prepared in the laboratory:

- Laboratory materials;
- Secondary reference materials;
- Control laboratory materials;
- Materials for quality control;
- Control samples.

For users of RMs, their position in the metrological hierarchy is important, but so is the utility aspect—what materials we have and what purposes they can be used for. Itemized are the non-matrix materials, meaning pure chemicals of specified high purity and matrix materials that may be synthetic (prepared by blending the ingredients in defined proportions) or natural (with the natural content of substance) or natural with added substance to be determined. A special group of materials is dedicated to a particular technique (analytical method) or to a specific sample preparation (e.g., extraction of metals from soil using aqua regia).

It is worth noting that sometimes the metrological quality (the positioning in the metrological hierarchy) of RM and its usefulness in a given measurement situation do not always match. In many textbooks, the metrological hierarchy of standards is given: the primary measuring standard is a pure substance, such as cholesterol; the secondary measuring standard would be cholesterol in the blood. In chemical measurements, the matrix material (secondary standard) can be more useful than standard of pure substance.

Below examples of RMs are listed, according to their use:

- Pure chemical substances of known purity (or known content of impurity) or their solutions commonly used as standards for calibration of measuring instruments (e.g., stock solution of cadmium nitrate used to prepare the series standards of cadmium for atomic absorption spectrometry), or to identify the presence of the substance (e.g., a solution containing nitrate (III) and nitrate (V) ions to confirm their presence in the water by ion chromatography).
- Pure chemical substances of known purity, used for reconstitution of the matrix (e.g., high purity copper of 99.99% Cu, used to prepare a series of standard solutions of zinc in the presence of copper for ICP-OES). Such materials should not contain the analyte determined in an analysis (e.g., copper with a purity of 99.99% does not contains zinc).
- Matrix RMs with the certified content of the substance (e.g., blood serum with certified cholesterol content).
- Matrix RMs used for the defined analytical procedure, reproducing the certified value provided the described procedures are used (e.g., fractionation of metals in soil with specified extraction media).

- Physico-chemical RMs for which properties such as density, viscosity and pH are certified (such materials are used for the calibration of measuring instruments, e.g., viscometers, pH meters).
- Materials with the assigned value of certain properties expressed in arbitrary units (e.g., fuel with the appropriate Research Octane Number (RON) or Motor Octane Number (MON); both numbers are evaluated with the same measurement process, the differences arise from the conditions used during the measurements).

6.4 Chemical Substances of a Given Purity

Substances of known chemical composition, its purity and available information on impurity content are primarily used for the calibration of measuring instruments and to identify the presence of the given substance in the test object. In practice, they are available as solids (in the form of a powder or compact solid) as gases (pure gas or gases mixture) or as liquids (prepared gravimetrically by weight a pure substance in a given volume of solvent). The uncertainty associated with the value describing the purity affects (it is a component of the budget uncertainty) the value of the overall uncertainty for the measurement procedure.

The chemical substances of high purity are also used to create the matrix composition corresponding to samples of the known main components. It is required however that the high purity substance do not contained the analyte (meaning that the analyte is below the detection limit of the used measuring technique).

6.5 Matrix Reference Materials

Materials prepared with the use of natural matters, such as soil, blood, leaves of plants, tissues of living organisms, dust, food products; these are just a few examples from a very extensive list of objects taken from nature. In terms of chemical measurements, matrix materials play an extremely important role, especially due to the fact that they behave chemically and physically the same as the respective test sample, assuming their identity or closeness to the chemical composition and physical properties of the investigated objects (samples). Matrix RMs should undergo processing that ensures their expected homogeneity and stability.

6.6 Applicability of Reference Materials

RMs may be used for different purposes. Those without certified values may be used to evaluate the precision of the measurements and in the quality control. Those RMs with certified values may be used, among other things, in:

– The development (optimization) of new measurement procedures;
– The process of validation of the measurement procedure;
– The verification of the measurement procedure;
– The evaluation of the accuracy of measurements;
– The calibration of the measuring instrument for a particular type of measurement. This allows the relationship between the received signal value and the content of the analyte in the sample to be determined;
– Proficiency testing: the use of a well-characterized material, with a reference value hidden from participants, allows the competence and proficiency of laboratories to be assessed in terms of the defined type of analysis, in relation to an independently determined reference value (rather than relative to the average value obtained on the basis of the results provided by the participants).

In all cases, the RM should be treated as a test sample for which measurements are made in the given laboratory conditions. Subsequently, the results (a value with assigned uncertainty) for the RM are compared with the certified value and its uncertainty. In everyday practice, quality terms are often used: stating that the results 'are consistent' or 'inconsistent' with the certified value. It is, however, recommended to apply a quantitative assessment of compliance, which involves comparison of the reference value with its assigned uncertainty of the mean value obtained in the laboratory and the respective uncertainty.

The certified value of a given property of CRM is to be used together with the uncertainty u_{CRM} stated on the certificate of CRM material. The absolute difference ΔC between the measured mean value of C and reference value C_{CRM} is calculated according to the formula:

$$\Delta C = |C_{\mathrm{CRM}} - C| \tag{6.1}$$

Uncertainties are usually expressed as standard deviations, and their variances are additive. Uncertainty of ΔC is thus $u\Delta$, which is calculated as the propagation of uncertainty of the reference value u_{CRM} and the uncertainty of the value determined in the laboratory u_C.

$$u_\Delta = \sqrt{u_{\mathrm{CRM}}^2 + u^2{}_c} \tag{6.2}$$

The expanded uncertainty U_Δ corresponding to a confidence interval of approximately 95% is obtained by multiplying the standard uncertainty by a coverage factor of $k = 2$.

To assess the accuracy of the measurement procedure used in the laboratory, the value of ΔC must be compared with U_Δ; if $\Delta C \le U_\Delta$, it is considered that there is no statistically significant difference between the measurement result and the reference value within the uncertainty.

In the case of the given CRM, for which several properties have associated reference values, the certificate contains expanded uncertainties values for each property. The certificate of the RM should also contain information on the procedure used for the evaluation of the uncertainty and information on the coverage factor k, applied. The standard uncertainty of the reference value is obtained by dividing the stated expanded uncertainty by that coverage factor.

In some cases, the uncertainty refers to 95% of the confidence interval of the mean value obtained with using the mean values delivered by laboratories participating in the evaluation of the reference value. In this case, the Student t-value for the 95% confidence level for n-1 degrees of freedom (n is the number of laboratories) is determined from the statistical tables for the t-distribution, which means that the coverage factor may not be equally 2. Then the standard uncertainty of the reference value is determined by dividing the expanded uncertainty by the coverage factor (varying from 2) specified in given certificate.

In many cases, the uncertainty assigned to the reference value is symmetrically allocated around this value. However, there are situations where the uncertainty is asymmetrically allocated. For example, for corn powder (ERM® BF418c; maize powder), the contents of genetically modified maize 1507 is given as 9.9 g/kg with assigned uncertainty of ($-0.6 \div +0.8$) g/kg. This means that the certified reference value has been assigned an asymmetric uncertainty range and the following approach should be applied when comparing the reference value with the value obtained in the laboratory:

- Positive uncertainty range (0.8 g/kg) should be used when the average value obtained in the laboratory is higher than the reference value;
- Negative uncertainty range (0.6 g/kg) should be used when the average value obtained in the laboratory is lower than the reference value.

6.7 Selection of Reference Materials

An important aspect of the laboratory practice is the proper selection of a suitable reference material, depending on the intended use. First, the commonsense principle, says that the best RM is one that meets the requirements regarding the use of the results of measurements for a given purpose.

An important criterion for assessing the suitability of a RM is to decide whether the material is to be used to calibrate the measuring instrument, or to assess the behavior of the analyte during the entire measurement procedure, including both the preparation of the test sample and the measurement itself. In the first case (calibration of measurement instrument), pure chemical substances are most frequently used, which allow the response of the measuring system (signal) to be assigned to a given quantity of the substance to be determined. The uncertainty of the reference value and the precision of the measurements a contribute to the uncertainty budget of the calibration process. In the second case (evaluation of the measurement procedure), the RM is used to evaluate bias of the results when using a selected measurement procedure, or the value (in percentage or as a fraction of) of the difference between the reference value assigned to the CRM and the value obtained in the laboratory using a given measurement procedure. A well-selected RM should reflect the behavior of the real samples, indicating that the content of determined substances, the content of the matrix, and the physical form of the reference should be the same as or close to the test sample.

The similarity (or the lack of it) of the RM to the test sample is an extremely important criterion for selection. For this, the term 'commutability' is used, defining the degree of conformity (similarity) of properties and behavior of the RM and sample in the measurement procedure.

Commutability: the property of a reference material, demonstrated by the equivalence of the mathematical relationships among the results of different measurement procedures for reference materials and for representative samples of the type intended to be measured.
Clause 2.1.20, ISO Guide 30:2015

In fact, the lack of full compliance of characteristics of the RM and test sample should be considered as a source of uncertainty, which should be then included in the uncertainty budget. It is not easy selecting the ideal RM that meets the requirements of full compliance. This is due to a number of constraints, which are explained below.

6.8 The Type of Reference Material Versus the Type of Sample

The RMPs offer RMs produced with the use of different kind of matrices; for example, water, soil or food products. At first glance, it seems that for the laboratory dealing with the examination of soil, an available reference material prepared from the soil would be straightforward. By examining this in more detail, it is essential to consider the kind of soil; for example, the content of organic matter or the content of silica in the soil from which the RM was prepared. Extraction processes work differently in the case of sandy soils, and otherwise in the case of soils containing substantial amounts of humic substances. Similarly, when purchasing a RM of milk powder, we must consider the content of fat; and, for geological materials, the presence of various minerals. Where the composition of the matrix of the sample and the RM is different, it can have a significant effect on the extraction process, and hence the effectiveness of transferring the substance to be examined into solution.

Another problem is the physical form of the RM. In many cases, manufacturers prepare materials in the form of a particulate powder of a high homogeneity. Such material is certainly very valuable from the point of view of its metrological quality, but, soil raw samples are after all not finely powdered and as well homogeneous as RMs. The question then arises whether the extraction process of the analyte from the soil taken from nature and that which undergoes specific processing so as to meet the requirement of the RM runs with the same efficiency. In practise it is not a case, thus this should be also considered, and if possible another RM (better fit to the original soil samples) should be taken. If this is not possible, the appropriate procedure for sample preparation of natural soil should be used (e.g., through its fragmentation to a form similar to the reference material). Another solution may be a recovery test carried out not only for a reference material but also for test samples.

In the case of clinical specimens, commonly available materials are usually in a lyophilized form (e.g., blood serum), as the test samples obtained from patients are usually delivered without being lyophilized. Of course, the freeze-dried material of the RM is subjected to a reconstruction of the liquid state by adding an appropriate volume of water, but certainly, afterwards the behavior of serum components could be different. The potential impact of the freeze-drying on the result should be considered.

A useful reference material should be as close as possible in terms of chemical composition to the matrix as well as in terms of the physical form of the test samples.

6.9 The Content of the Substance to Be Determined in the Reference Material and the Sample

The basic requirement is for the concentration of the substance in the reference material to be similar to the concentration that is present in the analyzed test samples. In practice, this is not always possible; often the available RMs have a much higher content of the substance and rarely much lower. Some manufacturers recommend the dilution of the prepared solution in such cases, but this will decrease not only the concentration of the substance to be determined but also the concentration of the matrix components, which can seriously affect the measurement result.

Another problem is the chemical form of the substance to be determined. This is due to two reasons: first, for different chemical forms, extraction efficiency may vary; second, in a number of measuring techniques, the detector response depends on the chemical form of the analyte. For example, for the determination of arsenic and selenium, by atomic absorption spectrometry with hydride generation, the efficiency of the reduction depends on the original oxidation state of the element (e.g., As(III) and As(V), and Se(IV) and Se(VI)). Another example is the determination of mercury by ICP MS (Inductively Coupled Plasma Mass Spectrometry); apart from a very high temperature of plasma, efficiency of the ionization of mercury present in the form of phenylmercury is significantly lower than the efficiency of ionization of the mercury present in the sample in the form of an inorganic ions, and this affects the number of ions reaching the mass detector. Such examples are numerous.

The appropriate reference material should contain the analyte at a concentration similar to its contents in the test samples, and in the same chemical form.

It is possible that natural materials are enriched by the addition of the analyte, so that its content is similar to that expected in the test samples. In this case, the form of the chemical substance added is extremely important. Enrichment of the material can be carried out in various ways. The simplest method is to add a known quantity of a substance to a prepared material (e.g., the addition of a standard solution of a mercury salt to serum; the addition of a standard solution of a pesticide to comminuted meat). In some cases, after the addition of a substance to the material, it is left for a certain time, assuming that it is possible to restore the chemical processes in the matrix (e.g., changes in a degree of oxidation; binding with protein).

For example, for the determination of total mercury content in blood serum, an additive of standard solution of $HgCl_2$ was used. It has been found that recovery is different depending on whether the measurements were carried out directly after the addition of the standard solution (almost 100% recovery was obtained) or if the sample was mixed under the incubation conditions before measurement (at the typical temperature of living organisms). In the latter case, recovery was only 80%. This suggests that incubation of the samples at around 40 °C has allowed the occurrence of biochemical processes that cause the incorporation of mercury with organic matrixes, therefore decreasing its ability for extraction.

An interesting example is the production of RM for testing the content of pesticides in foods of plant origin. In the case of tomatoes, it is possible to grow them in soil enriched with pesticides of interest or by adding to tomato pulp, dissolved in acetone mixtures of pesticides. One should therefore be aware of the method of preparation of the material, because it allows for the assessment of the possible impact of non-homogeneity of the material on the uncertainty of measurements.

It is worth noting that there are cases where the production of a RM is not possible; such an example may be where there is a need to determine volatile organic compounds in water. It is not possible to prepare sufficiently stable RM containing volatile organic compounds. In this case, the recommended solution is to produce mixtures of these compounds and add them directly to the test sample prior to measurement.

6.10 The Uncertainty of the Certified Value and the Expanded Uncertainty of the Measurement Result

In order to select the adequate CRM, it is also essential to consider the uncertainty associated to the reference values since this has an impact on the uncertainty of final result obtained for the test samples. The uncertainty attributed to the reference value depends on the type of material and the method used for the determination of the certified value. It was already noted that the position of the reference in the metrological hierarchy depends on the associated uncertainty. It is not always reasonable to purchase the material of the highest possible metrological quality, meaning with the lowest possible uncertainty of the certified value, due to economic aspects. The production of the RM and the certification process of the respective properties can be very expensive; thus, the price of the material depends strongly on the uncertainty with which the reference value is given. In practice, beyond the previously set requirements, the most reasonable and the economically justified choice is to look for the one whose standard uncertainty will not exceed the target combined uncertainty of the result. A good guideline is that the standard uncertainty assigned to the reference value does not exceed one third of the standard combined uncertainty for the result obtained with applied measurement procedure.

Example

The purpose of the laboratory is to determine the content of one of the most commonly used pesticide p,p'-DDT in soil samples. For this purpose, a standard procedure covered the extraction of p,p'-DDT followed by the GC-MS (Gas Chromatography Mass Spectrometry) measurements. In order to calibrate the instrument, laboratories use a pure reference substance. There are commercially available reagents with different purities, which have different prices.

Identification of kind of material	Declared purity	Contribution to uncertainty (%)	Approximate price (EUR)
Pure	>95%	5	~80
Pure for analysis	>99.0%	1	~150
Reference material	>99.6% ±0.4%	0.2	~230

Two measurement objectives:

1. In the first case, the purpose of the measurements is to evaluate which part of the 900 soil samples delivered to laboratory contains an elevated content of pesticide. Thus, a screening test will be performed, in which the accepted extended measurement uncertainty was set at 20%.
   Due to the requirement of 20% target uncertainty, it is economically reasonable to purchase substances with a purity of>95%, where the contribution to the uncertainty budget from the calibration of the measuring instrument (GC-MS) does not exceed 5%. According to the aforementioned rule, the standard uncertainty is less than one-third of the target uncertainty.
2. In the second case, the purpose of the measurements is the accurate determination of p,p'-DDE in these soil samples, where its elevated content was detected. The measurement results will be used for the decision regardless of whether the level specified in the directive as the maximum concentration was exceeded. The results of these measurements will be used to guide the possible punishment of producers using excessive amounts of pesticides for crops. In this case, the

accepted uncertainty of measurement was set at 1%, which means that it is necessary to purchase a RM with the lowest available uncertainty of the certified value (NIST gives 99.8% ± 0.2% for the RM 8467).

6.11 Proper Use of Reference Materials

The most important features of a RM should be its homogeneity and stability, and additionally, in the case of CRMs, the reference value with assigned uncertainty. Homogeneity should cover both the material in the single container, and the materials in various containers with the same batch. The stability of a material primarily concerns the stability of certain of its properties, and in the case of a CRM also refers to the stability of the certified value and its uncertainty.

The homogeneity of the material ensures that its subsequent portions taken from the container carry the same value for that property. Manufacturers must therefore determine the degree of homogeneity of the material, which in practice means that it is necessary to state in the certificate the smallest weighed portion of sample that guarantees the representativeness of the weighed portions of the material. The degree of homogeneity of the RM means that only using a single portion, no less than the manufacturer specifies as "the minimum amount of sample to be used," guarantees recovery of the reference value with assigned uncertainty.

The stability of the material depends on the type of matrix as well as on the environmental conditions used for its storage (e.g., temperature, humidity). The RM should be stable during the period of its validity, according to the manufacturer's statement. The producer should provide the required storage conditions of the material to guarantee the stated stability. These requirements typically involve the storage temperature of the material, for example, below −20 °C; in the range of from 1 to 5 °C; below 20 °C. They may also relate to humidity, especially for those materials for which an increase in humidity can change its properties. In the case of chemicals responsive to light radiation, storage of the material may also be required to be in a dark and/or light-tight container. The manufacturer may also provide special requirement for processing before using the material (e.g., drying under certain conditions, agitation).

Maintenance of the properties of the reference material is only guaranteed when the material is stored in the conditions specified by the producer.

The validity period given by RMPs guarantees the characteristics of the material at the first opening of the sealed container. After opening, the responsibility for the proper handling of the material falls on the user. It is important to ensure that, each time, only a portion of the necessary material is taken. In any case, the residues should not be re-introduced to the original packaging.

6.12 The Reference Value

In the case of CRMs, it is extremely important to pay attention to information on the aim of its use. The reference value specified in the certificate of RMs may be used independently from the procedure of sample preparation and/or the methods of measurement, or may be dependent on the measurement procedure. A good example is the set of RMs for the determination of metals in soil. It is known that, depending on the extraction medium used, the efficiency of the extraction varies for different chemical compounds in which the metal is present in the soil. In some cases, RMPs set a reference value assigned to a defined procedure for sample preparation. Therefore, sample preparation should be conducted in accordance with the described procedure guide to restore the value specified in the certificate. For example, in the documentation provided with "Soil CMI 7001 (light sandy soil)," three reference values are given for the determination of lead. These values (certified value in mg/kg and the assigned uncertainty in percentage) correspond respectively to the lead leaching: with nitric acid in room temperature (20.7 mg/kg of ±3%), with nitric acid on heating plate (23.7 mg/kg ± 6%) and using aqua regia (24.1 mg/kg ± 7%). This example shows the importance of a detailed analysis of the contents of the certificate of the purchased RM.

6.13 The Shelf Life of Reference Materials

The RMP is obliged to periodically study the material being sold, and if statistically significant changes are found for a given property, they are obliged to inform their clients (laboratories). It is expected that RMPs should provide the term of validity, which means that only in this period is the durability of specific properties assured.

There is a question as to whether it is possible to use a RM after its expiration date, and if so, under which conditions. A few years ago, it was a common opinion that no material could be used after that date and laboratories were discarding large amounts of very expensive chemicals. Currently, the commonsense approach prevails, where the quality control approach can be used to answer this question. It is recommended to regularly examine the RM used in the laboratory; thus, systematically collected information allows an assessment of the applicability of the material beyond its expiration date. If the reference value has not changed in terms of its assigned uncertainty, there is good reason to extend the validity of the material within the individual laboratory. In this case, it is necessary to carry out the evidence of confirming the stability of the reference value and to prepare a report confirming the extension of the validity of the material.

Recommendations of manufacturers of reference materials include:	
– The temperature of storage; – The minimum amount of sample to be used; – The hygroscopic feature of the material (the reference value may relate to the material dried at a predetermined temperature for a predetermined time); – The use of the certified value	

6.14 The Use of Certified Reference Materials for Ensuring the Traceability of the Result

The CRM ensures, the reference value with the assigned uncertainty held by this material. In many cases, the reference value is obtained from a statistical evaluation of results obtained in different laboratories. Unfortunately, obtaining very similar result by participating laboratories is no guarantee that the result is close to the true value, as the systematic effect cannot be detected.

It is worth remembering the meaning of the accuracy and precision of measurements; a precise result is not always an accurate result.

In chemical measurements, CRMs play a role similar to the standard units of the International System (SI) in a physical measurement, which means that they allow the transfer of properties (e.g., iron content in blood serum) between different laboratories and independent reproduction by the various entities. As a result, they constitute one of the most important tools to ensure measurement traceability, especially in those situations where it is not possible to establish direct reference to SI units. In this case, the principle is that the reference material of the highest possible metrological value is recognized by all interested parties. Thus, the periodic use of CRMs, together with the set of the test sample measurements, should provide measurement traceability.

The use of laboratory measurement standards and the comparison of these standards to a common standard of higher order (standard measurement reference) provides the traceability.

Suppose that two laboratories **LAB A** and **LAB B** conduct measurements of a given component in a given type of objects. Laboratory **LAB A** obtained a result

a and the laboratory **LAB B** obtained a result **b**. Both laboratories used the RMs (measurement standards) purchased from local producers **A** or **B**.

Consider two scenarios:

(1) Measurement standards A and B do not have a reference to a common standard, which causes a lack of measurement traceability in relation to a standard measurement reference.

Working measurement standard A → LAB A → result a

Working measurement standard B → LAB B → result b

In this case, it is NOT relevant to compare (meaning metrological comparison) the results obtained in both laboratories, due to the lack of a common reference.

(2) Measurement standards A and B have a reference to a common standard CRM, which ensures measurement traceability by reference to a common standard

CRM → working measurement standard A → LAB A → result a

CRM → working measurement standard B → LAB B → result b

In this case, it is relevant to compare (meaning metrological comparison) the results obtained in both laboratories, as both working standards are traceable to a common CRM.

The example discussed above shows the necessary condition for ensuring the metrological comparison of results from various laboratories. The generic meaning of the verb 'compare' differs to its use in the metrological sense. Thus, the metrological comparability means that results are expressed with the same unit of measurement and/or were obtained with the use of the same analytical procedure.

> The metrological comparability of measurement results: comparability of measurement results, for quantities of a given kind, which are metrologically traceable to the same reference.
> *Clause 2.46, ISO/IEC Guide 99*

The metrological comparability does not mean that the measured quantity values are of the same order of magnitude, but rather that they should be expressed in the same units.

The results that can be compared metrologically:

The results of the determination of zinc in two samples:

Sample A: 20 mg/L
Sample B: 120 mg/L
Conclusion: The content of zinc in sample B is six-fold higher compared with the contents of the zinc in sample A. The results can be compared.
Sample A: 20 mg/L
Sample B: 20 g/kg
Conclusion: Although the values are the same, the results are expressed in different units, so it is not possible to directly assess, in which sample the content of zinc is higher. The results cannot be compared.

6.15 Summary

There is no doubt that the use of RMs by laboratories is a crucial demand to ensure the quality of the results of chemical measurements. It is worth noting that RMs are often very expensive, so it often happens that laboratory management looks for opportunities to reduce costs by limiting the expenses relating to CRMs. Thus, the proper selection of the standards is essential to as to ensure the metrologically sound reference values with the acceptable uncertainty within the economically reasonable cost of the CRMs.

As previously emphasized several times, when choosing the appropriate RM one should primarily take into consideration its similarity to the test material in respect to the matrix, and in the content of the analyte. The uncertainty assigned to the reference value should be evaluated in respect to the expected uncertainty budget.

Each reference material should be accompanied by the appropriate document, often called a certificate or a certificate of reference material. According to ISO 17034, this document should include relevant information—namely, it must contain a reference value and its assigned uncertainty and information on the measurement traceability for the reference value (in the case of certified material). It is also expected that the manufacturer should give information on the aim of the use of the material and the storage conditions. In the case of chemical measurements, it is important to correctly determine the measured quantity. It is important to know whether the reference value is independent of the measurement procedure (e.g., determination of the element after the total digestion of soil samples), or is operationally defined (e.g., determination of the content of an element after the extraction of its water-soluble compounds). The RMP should clearly specify what the reference value refers to, and the laboratory should be thoroughly familiar with this information. Otherwise, you can expect that the laboratory will not be able to reproduce the expected reference value. There have been cases when the manufacturer gives the reference value in relation to the dried material at a suitable temperature, which can result from a hygroscopic nature of the substance. This means that the drying of the material under specified conditions should be included in the measurement procedure.

It is important to also have in mind the physical form of the RM; for example, in the case of powdered solids, it is the size of grains. The certificate should contain information about the minimum weight, which guarantees acceptable homogeneity of the substance. In this case, misplaced saving as well as retrieving smaller portions of the material can lead to erroneous results. As mentioned previously, it is important to store the material according to the producer's instructions.

In addition to the described issues relating to the proper use of RMs, it is also important to evaluate other properties of the used RMs. Several analytes are known to be present in the form of different chemical compounds, which is the objective of the study of chemical speciation. The use of RMs containing given element in various chemical forms can lead to incorrect results, especially where the detector response depends on the type of compound in which the element is present.

Before purchasing the material, it is worth checking whether the material has the status of a CRM or RM. Commercially available control materials are very useful in laboratory practice but do not have the status of a CRM.

It is also important to check whether the material is of natural origin or whether it has been prepared synthetically. One should also examine whether the analyte is present naturally or was added during the processing of the RM. It is also worth paying attention to the content of an analyte; often the content differs significantly from that expected in the sample tested in the laboratory.

From the description of the material, it is important to seek out information regarding the intended use of the material and how to ensure a reference value is obtained. Some CRMs are designed to recover the reference value through performing studies using the well-defined and specified analytical procedure. This can include sample preparation (e.g., the digestion of the soil or extraction using various media: aqua regia, hydrochloric acid, acetate buffer), and it may concern specific measurement techniques.

It is also important to known what the minimum amount of sample is for reproducing the reference value in terms of its value and uncertainty. Sometimes materials are sought for having superior homogeneity—in order to be able to apply as smallest aliquots. However, sometimes the most suitable materials are those mimic real samples, such as a grain size exceeding 1 mm and a homogeneity of no less than 1 g.

It is also important to evaluate how the certification is carried out. Frequently, certification is performed by statistical evaluation of the results from expert laboratories. By contrast, there are also materials for which the reference value was determined by the primary method (e.g., INAA or ID-MS).

Some materials requiring special storage are provided in appropriate packaging and under suitable conditions. If the certificate contains storage conditions that must be followed, the reference value is guaranteed only if the user follows the given recommendations; RMPs guarantee the stability of the material and the stability of the reference value during the validity of the material (shelf life). These properties are guaranteed until the opening of the package; thus, after opening, the user is required to use and store the material in accordance with guidelines provided. After opening, the packaging manufacturer is no longer responsible for the stability of the material. This does not mean automatically that the unused material (IMPORTANT: that remaing in the package!) is not suitable for use. The material is useful so long as it is kept under the appropriate conditions and used according to good laboratory practice.

Specified by the RMPs, the uncertainty of the reference values means that the value taken as certified value is in the range of assumed probability, usually 95%. The reference value for the properties, along with the assigned uncertainty, is taken as such when the CRM is used for calibration of the measuring instrument. If the CRM is used to evaluate the bias, the uncertainty is used for the evaluation whether it overlaps with the uncertainty associated with the value determined in the laboratory.

The structure of RM certificates (Table 6.1) is designed by the producer; however, various essential information should be always included—namely, certified value(s) accompanied by the uncertainty of properties in a given matrix (i.e. the type of sample).

Table 6.1 Important information that should be given in the reference material certificate

Information	Description	Comments
Name of the document	Certificate of reference material; Certificate of analysis	Depend on the producer
ID of CRM	CRM name, ID number	Crucial
Measurand	Name of chemical compound	Definition of measurand if it is method-dependent
Kind of object	Name of the material used	Type of the chemical matrix
Short description of material	Detailed information on the kind of material, including storage condition, safety issue	Note whether storage conditions refers to unopened container or after opening the container (due to e.g., moisture uptake or protective inert gas losses)
Certified value with uncertainty	Value accompanied by uncertainty at given probability (*k* factor)	Crucial
Minimum sample intake	Minimum amount of material that is representative for the whole CRM	Depends on homogeneity of material
Moisture determination	The conditions (temperature/time) used for determination of moisture content	Depends on how the certified value is given
Expire date of the certificate	Date related to the issue of certificate	Given by producer
Shelf life of the CRM	Information of the stability of CRM	Could be given as: – Final date (e.g., March 2019); – Period (e.g., 6 months after delivery or 2 months after opening); – Undefined, to the use of material
Signature of authorized person	Name, position and signature of authorized person	Given by producer

Chapter 7
Validation of the Measurement Procedure

Information versus disinformation

The primary objective of chemical measurements is to obtain **information** about the qualitative and quantitative composition of the object. The paramount requirement is, in this case, the reliability of the information so that the results are suitable for a given purpose of measurement. The results that are not correct (reliable) cause **disinformation** and consequently may lead to the wrong decisions being made. Therefore, the essential part of the of the selection of the analytical procedure is the unambiguous definition of the objective. In addition, it is also necessary to assess how the chosen measurement procedure is suited to the intended use of the results. In order to compare the commonly used feature of the analytical procedures, several parameters expressed in numerical form are used.

The purpose of measurement is to obtain reliable results, and this objective can be met when the analytical procedure used has been subjected to a process of validation. The primary task of validation is to ensure that the entire measuring process takes place as planned, and that the results are reliable and accurate.

Validation is the process of assessing the analytical performance of the measurement procedure and the compliance of the measurement procedure with the set requirements. The validation includes the determination of the relevant characteristics of the measurement procedure (e.g., accuracy, repeatability) and whether they are fit for the purpose of analyzing the laboratory's samples (e.g., recovery, uncertainty). The validation also includes the measurement instrumentation, the software of the instrument, the process of sample preparation, the process of data evaluation and the competences of the analyst (i.e. staff).

Validation is the process towards ensuring that the measuring procedure is suitable for the intended purpose, which also confirms that we know and understand the

E. Bulska, *Metrology in Chemistry*, Lecture Notes in Chemistry 101,
https://doi.org/10.1007/978-3-319-99206-8_7

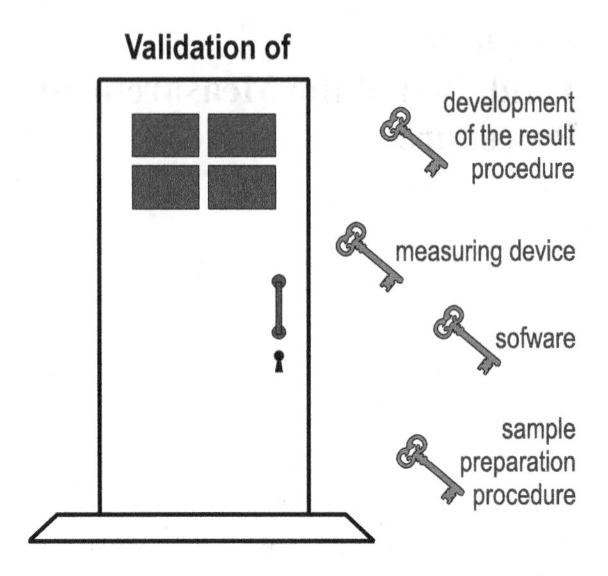

Fig. 7.1 Validation as a key for reliable results

established requirements, and we can confirm that the procedure used in the laboratory meets these requirements in terms of routine measurements. Validation is the key to reliable results (Fig. 7.1).

> Validation (from the Latin *validus*; meaning in English strong, powerful, worthy): documenting that a process or system meets its predetermined specifications and quality attributes.

Validation versus verification

The terms verification and validation have been described in the ISO/IEC 17025:2017, Clause 7.2. It is highlighted that validation includes specification of the requirements, determination of the characteristics of the methods, a check that the requirements can be fulfilled by using the method, and a statement on the validity.

Interestingly, in the current edition of the *International Vocabulary of Metrology—Basic and General Concepts and Associated Terms,* ISO/IEC Guide 99:2012, there are two definitions referring to the terms 'verification' and 'validation.'

> Verification: provision of objective evidence that a given item fulfils specified requirements*
> *Clause 2.44*

Validation: verification, where the specified requirements are adequate for an intended use**
Clause 2.45

In order to make those definitions suited to laboratory practice, several examples were added.

*This could be understood as confirmation that a given reference material as claimed is homogeneous for the quality value and measurement procedure concerned, down to a measurement portion with a mass of 10 mg.

**A measurement procedure, ordinarily used for the measurement of mass concentration of nitrogen in water, may also be validated for measurement in human serum.

The terms verification and validation and their definitions given in the VIM 3 both caused a lot of discussion among professionals dealing with measurements. First of all, it was essential to compare the definitions given in the VIM 3 with those on validation given in ISO/IEC 17025. Very briefly, the definition of the term 'verification' in VIM 3, "to provide objective evidence that an object meets specific requirements," covers the requirements set out in ISO/IEC 17025 for validation. The term 'verification' (or 'confirmation') was widely used in laboratories for the procedure used when confirming that analytical parameters of previously validated measurement procedures, for example, standardized procedures, can be achieved in a given laboratory environment. Owing to the extensive discussion around the distinction between the meaning of the two terms verification and validation, the following interpretation of the above definitions has been proposed:

Suppose a laboratory bought a new instrument, and after its installation, the laboratory planned a series of measurements to confirm whether the instrument fulfilled its specification. This process could be considered to be verification—the laboratory would be obtaining measurement data (objective evidence) to confirm the proper operation of the instrument. The next step would be to use the new instrument (which had previously undergone verification) with the analytical procedure, for which the laboratory identified the earlier analytical requirements, in line with those agreed with the customer, for the intended use of the procedure. As a result, the laboratory has planned a series of measurements aimed at obtaining confirmation that the measurement parameters are consistent with the requirements. This process could be considered as validation.

7.1 Techniques, Methods and Measurement Procedures

In the literature, various descriptions of the process used in chemical measurements can be found. In this work, terms are used that relate to the individual steps or rather

Table 7.1 Description of meaning of measurement technique, method and procedure

Description	Meaning and examples
Measurement technique	Analytical technique based on particular physical phenomena (e.g., atomic absorption spectrometry)
Measurement method	Detailed description of the method used for performing measurements (e.g., determination of calcium in mineral water by flame atomic absorption spectrometry/determination of lead in blood serum by electrothermal atomic absorption spectrometry)
Measurement (or analytical) procedure	Detailed description of the entire procedure (e.g., sample preparation, the list of reagents, their concentration, time of the certain processes, type of calibration)

a set of steps in the measurement process (Table 7.1). Therefore, the related terms 'technique,' 'method' and 'measurement procedure' can be distinguished from one another.

7.2 Validation of the Analytical Procedure

Validation of the measurement procedure covering all its steps is the most complex process that must be performed in order to document the analytical performance of the analytical procedure used for real samples. In this case, it is important to confirm that the method completes the requirements appropriate for its application. The scope of validation, or range of parameters to be set in order to characterize the procedure used depends on whether the validation process is subjected to a standardized or laboratory method.

Chemical measurements are used to characterize the object studied, which clearly implies the need to deal with real samples. In the ISO 9000:2000, the definition of the term 'validation' is accompanied by a note stating that validation can be conducted under real or simulated conditions. In the case of chemical measurements, validation should be carried out not only with the use of pure standard solutions but should also consider the complexity and variability of the investigated objects. The simulation should take into account the conditions of use of the measurement procedure and the nature of samples. It is also wise to consider the use of the results as well as problems arising from the users of those results.

In some cases, however, it is helpful to define the primary parameters of the analytical measurement method, giving evidence about the measurement capabilities of the given procedure used at a laboratory. Thus, it is reasonable to carry out analytical processes for model systems. However, it is then necessary to ensure that the use of a given procedure for a customer's specific problem is suitable for the intended purpose.

7.3 Verification and Validation of Measuring Device

Measuring equipment, before being placed into service, should be checked to establish that it meets the laboratory's specification requirements. It should be therefore subjected to formal acceptance, meaning that it should be entered into the record, be assigned the appropriate identification number, and assigned the adequate metrology status (i.e., calibration, testing). In technical terms, the equipment should be checked for its technical specifications, compliance with the environmental requirements of the manufacturer, and then calibrated to its assigned metrological status. In order to performed the primary examination of the given equipment, tests are carried out by using pure chemicals in most cases. This allows the performance of the device to be monitored under standardized conditions. It is of course expected that the equipment should be operated by qualified and authorized personnel.

Once enrolled, the device should undergo monitoring of its performance and stability—thus, the maintenance plan should be created. The basic criterion for establishing the frequency of the check of the performance should be that intervals are shorter than the time in which the deviation of key parameters can go beyond acceptable limits (which can be evaluated on the basis of laboratory experience or risk assessment).

7.4 Validation of the Measurement Device Software

The validation of software used for controlling the system should be the task of the manufacturer. The laboratory may require the supplier to deliver a validation report; moreover, the manufacturer should provide assistance in the periodic testing of the software, or of providing test packets, allowing the user to independently verify the operation of the system according to the specification. In practice, many modern devices have inbuilt package tests, which check the computer system each time after starting.

A common practice of manufacturers is updating the software that controls the device; thus, revalidation is required in the case of any modification or change. Updating the software may affect the mode of operation of both the computer and the measuring device, so it is necessary to have evidence, indicating that the updated version of the software has undergone a process of re-validation.

7.5 Validation of Procedures for Processing the Results

In order to obtain the final results, first the calibration of the given equipment should be done; then the measurement for the set of test samples can be performed, resulting in raw data (signals) that can then be subjected to processing. In practice, raw data collected are subjected to various mathematical procedures, which use different algorithms. This applies, for example, to using the various tests for rejecting outliers, various classifying tests, whether the results belong to the same population, different algorithms depending on the determination of the calibration, as well as various methods of estimating uncertainty. The primary criterion is that for the given type of measurements, the same algorithms should always be used. Moreover, the analyst must prove that the algorithm is correct for a particular measurement procedure. This requirement stems from the fact that, depending on the intended purpose of the research, there are different methods of calibration, a different number of repetitions, and differences depending on the detector response to the actual sample composition. Thus, the calculations involved must be selected individually, and the legitimacy of the choice must be confirmed in the validation process.

In practice, the term 'software validation,' is used, which refers, for example, to spreadsheets. In this case, it should be considered as a mental shortcut. The user of the spreadsheet does not need to validate the software as such, provided that the software is delivered by a renowned manufacturer or a source for which the manufacturer has supplied a suitable report confirming that the program meets certain requirements (verification).

The above-mentioned term applies in practice to the validation of algorithms included in a spreadsheet and used for the conversion of the inserted raw data. In commercially available spreadsheets, there are algorithms for rounding numbers, and for executing multiple operations that may affect the final result of the calculation. In practice, so-called 'software validation' is carried out by performing a model calculation (for the prepared set of data), both using the software program as well as another program or a calculator. The comparability of both results within the range of accepted uncertainty is considered to be confirmation that the requirements for the intended use have been met.

7.6 Standardized Versus Laboratory-Developed Procedures

It is clear that the analyst should use analytical methods and measurement procedures appropriate for the expected application. According to the requirements of ISO/IEC 17025:2005, methods published in the appropriate international, regional or national standards or by reputable technical organizations are preferred. This is mainly due to the fact that standardized methods have undergone the laborious process of validation by expert laboratories and are considered optimal for a particular purpose. Moreover, the use of the same analytical procedures in laboratories conducting the same kind of testing (e.g., the monitoring of surface water pollution, food inspection, clinical trials), is the right basis for the metrological comparison of results, especially when they are dependent on the applied measurement procedure. However, it is important to note that the use of a standardized method does automatically ensure the correct result; the laboratory should confirm (verify) that their conditions are appropriate for conducting the assay. Although in this case there is no formal requirement to perform validation, in practice, selected parameters are used for confirmation of validation. The choice of these parameters depends on the purpose of the testing.

In the case of environmental studies, one can expect high variability of the composition of soil (e.g., natural sample versus samples from contaminated sites), which means that the critical parameter will be the calibration range.

In the case of doping control, it is important to detect whether the prohibited substances are present—hence the critical parameter of evaluation of the measurement procedure will be the limit of detection and uncertainty.

The use of standard methods has many advantages, unless the analytical parameters of the method and its application area comply with the requirements for the intended use. However, there are some situations where the standard method is to be used outside of the validated range, or the laboratory deems it appropriate to modify the standard analytical procedure. In such cases, it is necessary, according to the formal requirements, to conduct validation. The scope of the validation should be appropriate to the specific requirements in relation to the analytical parameters.

Typically, the process of establishing standards is laborious, while in turn the development of measurement techniques is nowadays extremely fast. Thus, it may be the case that the given standard is outdated. In such cases, the laboratory may have its own analytical procedure responding to the demand. This is also valid when there is no specific standard describing the particular kind of testing. In both cases, it is necessary to conduct validation.

Measurement procedures requiring validation

- Non-standardized methods developed in laboratory;
- Non-standardized methods described in the scientific literature;
- Standardized methods used outside their scope;
- Standardized methods modified in order to be fit for a purpose.

Measurement procedures requiring confirmation

- Standardized methods.

It should be noted that the purpose of validation is not, in any case, assessment of the analyst performing the determination, or evaluation of the laboratory. Validation is primarily aimed at determining the analytical performance of the measurement procedure in order to decide on its suitability for the intended use. Thus, validation is a useful tool for evaluating the quality of the results, especially with regard to trade, and the legal context, as well as in scientific research.

Validation is confirmation by examination and the provision of objective evidence that the particular requirements for a specific intended use have been fulfilled.
Clause 5.4.5.1.ISO/IEC 17025:2005

7.7 Measurement Techniques

Before the introduction of analytical instrumental techniques, analytical methods using chemical reactions were commonly used. The course of the reaction was observed visually or determined based on direct measurements of characteristics such as mass, volume or current. For example, determination of the amount of sulphate present was performed by weighting the precipitate of low soluble barium sulphate; the more sulphates there were in the solution, the greater the mass of the precipitate. For the determination of chloride by the coulometric method: the more chloride ions there were in the solution, the greater the measured electric charge. The dependence of the response of the measuring instrument (such as balance or voltameter) from the amount of substance (e.g., sulphate, chloride ions) was directly proportional only for the model systems containing the only determined substance. A more complex situation occurs in the case of real samples with a complex composition. In this case, it is important to evaluate the selectivity of the applied method in respect to a given property. The measurement of the mass of the substance is nonselective, which means that if the conditions of the precipitation of barium sulphate cause precipitation of the other compounds, the weighted mass includes all of the precipitated substance—meaning that it is not possible to distinguished individual components. Similarly, in the coulometric measurements, if other compounds that are carriers of charge are present, the coulometer will record the electrical signal (the number of coulombs) that corresponds to the content of all compounds undergoing an electrochemical reaction. These two simple examples indicate clearly that in some cases the chemical measurements do not only depend on the amount of substance of interest but may also depend on the composition of the object. These effects are the reason that in the case of real samples, attention should be paid to the proper selection of the conditions in which the test is conducted.

The rapid development of measurement techniques has meant that contemporary analytical chemistry uses various physical and chemical processes, the effects of which can cause various phenomena, which in turn can be measured—as long as a suitable measuring system is available. In spectral methods, various effects of the electromagnetic radiation within a wide wavelength range can be measured. There are also phenomena occurring during the exposure of the sample with a stream of electrons, ions, or uncharged particles. The electrochemical methods use electric quantities for measuring electroactive substances. In addition, immunological and enzymatic processes are being used more often, to follow the biochemical and biological compounds in particular.

An important part of the analytical process is the preparation of the sample prior to testing. The most commonly used methods of preparation are digestion or selective extraction; in many cases, chromatographic techniques are used, which allows components present in the sample to be separated. These processes may already be largely automated.

The development of instrumental methods and the increasing availability of advanced measuring instruments enabling computer control of the measurement process and the ability to automatically obtain a result in the defined units mean that the role of the analyst is often seen in the context of the skills required to operate specific equipment. This is due, among other things, to the fact that modern devices are equipped with detailed instructions and regulations for how to performed specific tests. Manufacturers outdo in praising their instruments as user-friendly and not requiring advanced knowledge (Table 7.2).

Such promises give the impression that the work of the analytical chemist is only to provide technical support to these great and universal measuring instruments. This would mean that chemical analysis would be limited to measuring the analytical signal and comparing it with the previously prepared calibration—not taking into account the effect of other components on the detector response or the stability of the measuring instrument. Such an approach, in isolation from the whole analytical process, could not guarantee a reliable result.

Table 7.2 Selected statements from the manufacturers' brochures

Example	From producer's specification
1	Our company Y has introduced a device X of unprecedented measurement capabilities. The device is designed for the most demanding applications
2	To meet the needs of the life sciences, company Y launched the device X giving unlimited possibilities in terms of measurement
3	Our most modern measuring instrument offers dozens of ready-made calibration curves, and simultaneously it is equipped with a validation packet
4	Your laboratory every day faces increasingly difficult challenges. Now the family of laboratory systems X of company Y gives new opportunities to meet these demands. The use of the instrument X ensures that even the most complex determinations become easy and routine

7.8 Validation of Measurement Procedure

Analysts are primarily interested in real samples, whose qualitative and quantitative composition is very complex, and what is more—unknown. Thus, the evaluation of the measuring procedure cannot be limited only to the measuring technique, but should cover all stages, including the collection and preparation of samples. The analyst is not only responsible for the correct implementation of a given measurement technique for determining the content of a substance in a complex object, but should also understand the problem of a customer and be aware on how the results will be used—this will involve the proper definition of the measurand. As a result, more and more attention is paid to all stages of the measurement process and their impact on the final result. This is related to an in-depth understanding of the variability of chemical equilibria and the changes the analyzed objects undergo within the storage and preparation process.

Validation of the measurement procedure is considered to be one of the most important responsibilities of the laboratory. When selecting an appropriate analytical method, the laboratory should take into account their experience and the capabilities of their infrastructure, above all else; the laboratory should also consider the time and cost of analysis. Validation of the selected procedure allows the customer to evaluate relevant evidence documenting the analytical parameters of the proposed procedure.

The scheme of the validation process:

- To identify all requirements for the intended use of the results;
- To determine essential analytical parameters of the proposed measurement procedure;
- To compare the obtained parameters with the requirements;
- To confirm that the selected measuring procedure meets the predetermined requirements.

In this aspect, the analytical process is multifacted, and the most important aspects may include sampling a representative portion of the test object; the preparation of an appropriate number of analytical sub-samples, so that it is possible to make repetitive measurements; the preparation of laboratory samples for testing (which includes dissolution, digestion, extraction, concentration, dilution, separation, etc.); calibration of the measuring instrument; monitoring the reliability of the results; and the evaluation of the results obtained.

As validation requires confirmation that the specific requirements were met for the specific intended use, the most essential part is the assessment of the analytical performance of the measurement procedure and the evaluation of selected features of the result. The parameters characterizing the measurement procedure are measurement range, linearity, sensitivity, limits of detection and quantitation. Moreover, there are parameters that characterize the result of the determination—namely, measurement traceability and uncertainty of the results (Table 7.3).

The laboratory may use the various tools enabling validation, those described in the relevant standards or by the authorized institutions. In the case of reference

Table 7.3 Analytical parameters of measurement procedure and results

Analytical parameters of measurement procedure	
– Measurement range – Limit of detection/quantitation – Sensitivity (slope of the calibration) – Robustness	
Feature of the analytical results obtained with the use of a given measurement procedure	
– Metrological traceability	
– Uncertainty of the results	Parameters influencing the uncertainty – Recovery – Robustness – Selectivity – Specificity – Repeatability – Reproducibility

laboratories, it is recommended to perform so called 'step-by-step,' validation, as described in GUM (*ISO Guide to the Expression of Uncertainty in Measurement*). This approach involves a systematic evaluation of all quantities affecting the measurement result. Thus, it is clear that the validation and evaluation of uncertainty of the result should always be considered together.

The techniques used for the determination of the performance of a method to be used for a given purpose are as follows:

- Calibration using reference standards or reference materials;
- Comparison of results obtained with another method;
- Interlaboratory comparisons;
- Systematic assessment of the factors influencing the result;
- Assessment of the uncertainty of the results.

7.9 Optimization and Validation of the Measurement Procedure

The measurement procedure, unless there is a standardized one, is subject to a preliminary optimization, in order to choose the experimental conditions that will allow the best performance and meet the requirements. For example, the resolution of the chromatographic separation depends on the composition of the eluent and its flow rate; the extraction efficiency often depends on the pH; the atomization process in flame atomic absorption spectrometry depends on the kind and flow of flame gases; and the number of counts in the mass spectrometry depends on the voltage applied to

Table 7.4 General requirements for the scope and range of validation

Range	To be considered
The entire measurement procedure	All the steps: from the sampling of an aliquot of the sample through its preparation to the measurement signal and the calculation of result accompanied with its uncertainty
Matrix variability	The variability of the composition of the matrix: for example, the varying content of organic matter in soils; the salinity of water; the sugar content in fruit
Expected content of analyte	The range of content of the analyte in test samples

the quadrupole analyzer. Optimization of the experimental conditions in such cases involves an assessment of the detector response at varying values of the parameters in the selected range (e.g., checking the absorbance of the standard solution of iron with a flow of carrier gas in the range of 8–12 mL/min). The assessment criterion depends on the purpose of the study, but it is usually sensitivity and precision. Optimization of the measurement parameters is followed by its further validation. In any case, it is important to conduct optimization/validation for the entire range of matrices and analyte concentrations in which the procedure can be used (Table 7.4).

Within the validation process, the main aim of the analyst is to collect all the necessary measurement data, so as to show that the selected measuring procedure can be used for its intended task, which meets the established requirements. The validation process must, therefore, cover the widest possible range of variables in order to comply with the analytical requirements. However, this does not mean that all the possible parameters should be tested each time. One should focus on those that are relevant to the type of test and to the intended use of the results. It is important to take into account a commonsense consideration of the use of the results and the necessary costs associated with conducting validation. After all the data for validation are collected, one should remember to carry out a formal 'statement validation,' which confirms that the applied test method is justified for the particular purpose. Such confirmation may be prepared in the form of a report or any other document.

The validation shall be as extensive as is necessary to meet the needs of given application or field of application.
Follows Clause 7.2.2.1, ISO/IEC 17025:2017

Example 1 Determination of the content of cholesterol in human serum.

Suppose the client (e.g., an institution responsible for public health) is interested in information on the number of persons for whom cholesterol content is close to the limit values above which medical interventions are necessary. The typical content of cholesterol is far above the detection limit of commonly used procedures in clinical labs.

Requirements for measuring procedure: the limit of detection is not as important, but the value of **uncertainty of the result** is important, which means that this should be evaluated carefully within validation.

Example 2 Monitoring of the emergency substances in an environment.

Suppose the client (e.g., a company dealing with soil phytoremediation) wants to know the most contaminated places on the site; the rough and fast screening procedure would then be required.

Requirements for the measurement procedure: the detection limit and uncertainty of the result obtained are not critical parameters, whereas the working range (linear) of the measurement procedure is important, which means that that this should be evaluated carefully within validation.

Example 3 Doping control in sport.

Suppose the client (e.g. the anti-doping organization) is interested in whether the presence of a prohibited chemical compound can be detected in the urine of an athlete participating in a competition.

Requirements for measurement procedure: the detection limit and uncertainty of the result parameters are extremely important, which means that this should be evaluated carefully within validation. In this case, a large linear range of calibration is not critical.

Both testing and calibration laboratories should estimate the uncertainty of measurement according to their own implemented procedures. They need to identify all the components of uncertainty and make a reasonable estimation based on knowledge of the performance of the applied method and previous experience and validation data. Thus, laboratories should be aware of the uncertainty of measurements.

The test report should at least include the test results with, where appropriate, the units of measurement. The uncertainty value should be added on request of customer, which should be always include as a part of review of the request and contract with a given customer.

In terms of the calibration certificate, they should include the uncertainty of measurement and/or a statement of compliance with an identified metrological specification.

7.10 Criteria for Selecting the Measurement Procedure

Before selecting the most appropriate measurement procedure, tailor-made for the intended research the analyst should participate in a discussion on the aim of performing measurements. Although the terms of reference should be established by client, the analyst should be aware of the use of the result. Formally, customers of the laboratory can be of external origin or internal origin, the latter meaning from the same organization. In both cases, it is essential to evaluate whether the laboratory has the necessary resources, including the competent personnel to implement ordered tests or calibrations. Arrangements can be of a more or less formal nature; the less formal approach applies primarily to internal clients, for which the contract review can be performed in a simplified way. It is worth remembering that the 'client' may also be the analyst; this applies to those cases in which research studies requiring measurements are conducted, for example.

It is relevant to recall the part of Clause 5.4.2 (ISO/IEC 17025:2005), stating that, "The laboratory shall use test and/or calibration methods, including methods for sampling, which meet the needs of the customer and which are appropriate for the test and/or calibrations it undertakes."

Therefore, it is clear that the analysts should orient themselves in issues related to the conduct of measurements, which allows for the evaluation of the problem from the point of view of a particular choice of the analytical process. The responsibilities of the analyst include discussing with the client the possible measurement capability, so as to avoid improper actions or erroneous conclusions.

The selection of the measurement procedure requires a decision stemming from the knowledge of the object (type of matrix and analyte) as well as the intended use of the results. Thus, the analyst must collect and evaluate a range of information relating to the expected testing and quality of results (Table 7.5).

7.11 Validation Parameters

Validation of the measurement procedure can be carried out within the laboratory (*single laboratory validation*) and by interlaboratory comparisons in which the participating laboratories use the same procedure. In any case, validation includes the specification of analytical requirements; the evaluation of selected analytical parameters; assessment of the extent to which the expected requirements can be met; and finally the preparation of the report, with a clear statement about the suitability of the procedures for the intended purpose (the validation claim). Measures of this assessment are assigned numerical values for the individual parameters. The most important parameters are selectivity, the range of calibration, sensitivity, limit of detection and determination, robustness, accuracy and precision.

Selectivity defines the possibility of determining a compound of interest in the presence of other components in the sample (i.e. the matrix). A method that is perfectly

Table 7.5 Check list for the characterization of the analytical task

Item	Description
Analyte	Define the kind of substance to be measured (e.g., iron)
Measurand	Define the analyte and the type of object to be analyzed (e.g., the content of iron in human blood serum)
Type of measurements	Specifically define the chemical species to be measured in the given type of object to be analyzed (e.g., the content of iron binded with proteins in human blood serum)
Range of analyte content	Predict the expected range of the content of analyte in test objects
Diversity of the matrix composition	Predict the composition of the object and the range of their diversity for test objects
General characteristic of test objects	Ascertain the state of matter (e.g. solid, gaseous); main components; traces
Availability of samples	Available mass or volume of a sample; necessity of performing nondestructive testing or clearance to destroy the sample
Preliminary sampling of the object	Who and how will perform sampling on side; transport and storage of samples?
Homogeneity of samples	What is the minimum sample mass/volume that is representative in respect to examined characteristics
Feathers of the results	Ascertain requirements related to accuracy and uncertainty—both from the clients and/or from legal regulations

selective for the determining the analyte is said to be 'specific.' The selectivity of a given method should be evaluated by performing measurements in the presence of matrix components, from the measurement of a matrix-free sample to matrix-rich mixtures representing the most complex samples. All observed influences of other components (i.e. interference) should be described in detail in the method documentation.

The range of calibration is a parameter characterizing the flow of the calibration graph over the range necessary to analyze samples of varying concentrations of analyte. Within validation, it is important to examine extent to which the required accuracy and precision is obtained. It is recommended that the concentration of analyte in a sample lies within the linear range of the calibration. In case this is not possible, one can apply a suitable non-linear algorithm. In practice, in order to establish the calibration curve, usually, five standard solutions and a blank are prepared. With significant deviations from linearity, it can be necessary to prepare more standard solutions. It is then advised that full characterization of the calibration (e.g., number of standards, slope, regression line) be included in the validation report.

The sensitivity of the measuring system is understand as the ration of change of the signal and the corresponding change in measured values (Clause 4.12, VIM 3). The definition of sensitivity given in VIM 3 is supplemented with two notes: (1) sensitivity of the measuring system can depend on the value of the quantity being measured; and (2) the change considered in a value of a quantity being measured must be large compared with the resolution. The sensitivity may be represented as the slope of the calibration curve, which may be described by the least squares mathematical function or calculated from the signal values for the two solutions of different concentrations of analyte.

Detection limit determines the amount/concentration of analyte that corresponds to the signal calculated from the value of the blank plus three times the standard deviation at a level close to the blank sample. This is the signal that, with a certain probability, can be distinguished from the blank.

The definition of the term 'limit of detection' is also given in VIM 3 (Clause 4.18), according to which it is the measured quantity value, obtained by a given measurement procedure, for which the probability of falsely claiming the absence of a component in a material is β, given a probability $\acute{\alpha}$ of a falsely claiming its presence.

The definition is supplemented with three notes: (1) IUPAC recommends default values for $\acute{\alpha}$ and β equal to 0.05; (2) The abbreviation LOD is sometimes used; (3) The term 'sensitivity' is discouraged for 'detection limit.'

Quantitation limit is the lowest amount/concentration of the analyte that may be determined by the given measuring procedure with the specified accuracy and precision. The value of the limit of quantitation should be evaluated by appropriate standard solutions. In practice, the solution of concentration corresponding to the detection limit is the lowest, except the blank sample, a point on the calibration curve.

Ruggedness/robustness, sometimes referred to as method tolerance, is a parameter characterizing the effect of small changes in procedures on the stability of the obtained analytical results. When using the method by different laboratories, as well as in the same laboratory over the time, one can expect some minor variations to occur. During method validation, it is important to determine which parameters of the analytical procedure are most influenced by other factors—meaning which are the most critical steps. Detailed information about how to validate the robustness of the method are given later in the text.

Accuracy is defined in the VIM 3 (Clause 2.13) as the "closeness of agreement between a measured quantity value and a true quantity value of a measurand." Attention should be given to the fact that the concept of 'measurement accuracy' is not of qualitative nature and is not expressed in a numerical quantity value. The smaller the measurement error, the more accurate the measurement.

Trueness, understood as correctness of the measurement was defined in the VIM 3 (Clause 2.14) as the closeness of agreement between the average of an infinite number of replicate measured quantity values and the reference quantity values.

Warning! The terms, 'accuracy' and 'trueness' should not be used interchangeably.

Table 7.6 Example list of parameters with the expected scope of validation

Parameter	Trace analysis	Major component analysis
Accuracy	X	X
Precision	X	X
Linear range	X	X
Selectivity	X	X
Detection/quantitation limits	–	–
Robustness	X	X

Precision of measurement is defined in VIM 3 (Clause 2.15) as closeness of agreement between indications or measured quantity values obtained by replicate measurements on the same or similar objects under specific conditions.

In practice, it can be stated that the precision characterizes the scattering of the results or series of measurements for a given sample using the analytical method. The precision is usually expressed as a standard deviation for the set of data. The precision of the method depends largely on the concentration of the analyte and this should be described clearly in the documentation of method validation. ***Repeatability*** refers to the results obtained under the same conditions, or in the same laboratory, using the chosen method for the sample by the analyst in a short period of time. ***Intermediate precision*** (also called the intra-laboratory reproducibility) refers to the results obtained in the laboratory over a longer period of time, even when studies are conducted by more than one analyst. ***Inter-laboratory reproducibility*** relates to the results obtained under the same measurement conditions, that is, the same method in another laboratory using another instrument by another analyst, over a longer period of time.

The scope of validation is the set of parameters evaluated within validation; this depends on the purpose of test, the type of analyte and its concentration. For example, in the testing of natural waters, the scope of the validation can vary for trace and major substances (Table 7.6).

Validation should be performed for the measurement procedure used routinely in the laboratory for real samples of complex composition.

7.12 The Frequency of Revalidation

There are no explicit requirements in respect to the scope and frequency of conducting validation of an analytical procedure used in the laboratory. It is clear, however, that once-in-a-lifetime validation cannot justify the use of a procedure for a longer period of time. In analytical practice, it is recommended that validation should be carried out again after each change of conditions. If so the validation within the originally applied full scope or if justified with the limited scope, should be done, ensuring that the procedure become under control (Table 7.7).

The scope of analytical parameters to be evaluated in the process of validation and/or re-validation also depends on the kind of testing (qualitative or quantitative analysis; unique testing conducted for one or only few samples; routine testing of a large number of samples, delivered to the laboratory over long time), the requirements for quality performance (accuracy, precision, uncertainty); and the time required for completion. The more parameters we take into account in the process of validation, the more time-consuming and cost-intensive it is to carry out the validation. However, the greater the demand on the quality of the results, the more often the process of revalidation should be performed.

Validation is always a balance between costs, risks and technical possibilities.

IMPORTANT! A significant source of information in the validation process, especially during revalidation, are data collected during quality control.

Table 7.7 Requirements for re-validation

Modification	What should be revalidated
A new kind of sample	Internal standard samples, standard addition, the measurement run for two or more portions of the sample
A new series of samples	Blank sample, re-calibration, reference materials
A new instrument	Check measurement parameters: measurement precision, limit of detection and quantitation, verification of analytical parameters for standard samples (internal or provided by the manufacturer)
A new bath of reagents	Specification/blank samples
A new standard	Comparison with a known standard solution, comparing the internal references
A new kind of matrix	Interlaboratory comparisons, a suitable reference
New personnel	Testing the precision, linearity, detection and quantitation limits, within laboratory comparison

7.13 Selected Parameters of Validation of the Measurement Procedure

This chapter describes in detail how to evaluate the characteristic parameters of the measuring procedure.

> Warning! In this section, information on the evaluation of the individual parameters must be considered as guidelines rather than as definite recommendations.
>
> For example, in many situations, the use of "at least 10 repetitions" is not possible and/or economically justified. Common sense should be utilized.

7.14 Selectivity and Specificity

Selectivity and specificity of the measurement procedure are terms often used as synonymous. Yet these terms are not synonymous; selectivity is the degree to which other substances present in the sample affect the analytical signal, while specificity is the ability to measure only the substance to be determined, without any interference from other components. The concept of selectivity is related to the assessment of how the received analytical signal is exclusively due to the presence of the analyte, and not to the presence of other substances with similar physical or chemical characteristics.

Selectivity characterizes, therefore, the degree of disturbance of the measured signal due to other substances present in the sample or other kinds of interferences. It should be stressed that the most difficult situation occurs if the effect of potential interfering substances is unknown.

Selectivity is a qualitative parameter, which may be graded or expressed descriptively; for example, it may be specified that the determination of selenium by ICP MS, based on the measurement signal for the most abundant stable isotope ^{80}Se, is selective in the presence of zwitterion argon containing an isotope of argon of mass 40 amu ($^{80}Ar_2$).

> Assessment of selectivity
> - Testing of the matrix-free samples and those contains interfering substances;
> - Evaluation of recovery for CRMs;
> - Comparing the results for different analytical methods.

7.15 Accuracy and Trueness

According to the definition given earlier, accuracy refers to the true value of a quantity, which means that without knowing it, it is not possible to evaluate the accuracy of the measurement. In practice, as the knowledge of true value is not accessed, one should use the term 'trueness' in direct association with a reference value and a measurement error.

- ***Trueness***: how close to the reference value is to the average of the series of measurements for a given material. Trueness is described by the systematic error of measurement, which can be expressed as an absolute or relative error.
- ***The absolute error***: the difference between the result of the measurement (usually this is the average value of a series of results obtained under repeatability conditions) and the reference value. Absolute error is expressed in units of concentration.
- ***Relative error***: proportion of the absolute error value to the reference value; this can be expressed as a dimensionless value or a percentage.

Assessment of the trueness of the analytical results

- Measure simultaneously the content of analyte in CRM and in the blank sample. Measurements should be performed with at least several repetitions (10 repetitions minimum is recommended*);
- Subtract the average value for a blank from the average value for the CRM;
- Calculate the standard deviation of the mean values for both;
- Calculate the standard deviation of determined analyte content (apply the law of propagation);
- Compare the reference value with the laboratory result (visually or with the t-test).

*The recommended number of repetitive values allows the use of sound statistical evaluation. However, one should be aware that there are situations when this is not possible due to the lack of a sufficient amount of sample or the high cost of a single measurement

The numerical value of the relative error depends on the content of an analyte; in most cases, the lower the concentration, the greater the relative error. Although it depends on the kind of analyte and matrix, the general tendency of most commonly observed values can be exemplified visually, as presented on Fig. 7.2. This however should be considered as indicative—regarded as typical for a given range of concentrations. Needless to state, that in specific cases, the relative error can diametrically differ from those indicated in the illustration.

Precision is the degree of compliance of measurements values for the series of repetitions, or spread of results around the average value. Precision is described by the standard deviation, the relative standard deviation, confidence interval or range.

The value of the relative standard deviation depends on the content of an analyte—the lower the concentration, the lower the precision can be. Although it depends on the kind of analyte and matrix, the general tendency of most commonly observed values can be exemplified by values listed in Table 7.8. This however should be

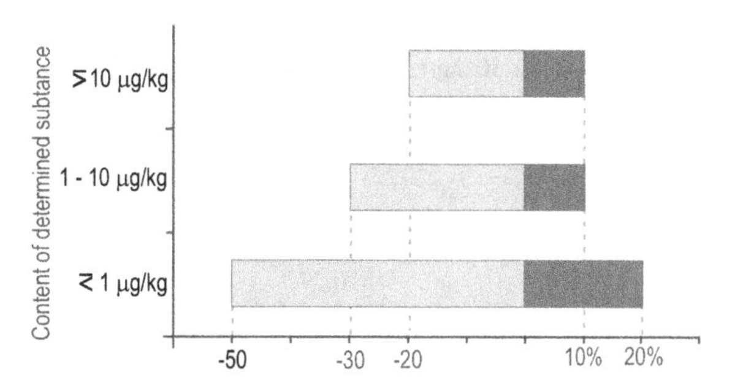

Fig. 7.2 General guidelines for the typical recovery depending on the concentration

Assessment of precision
For the calibration curve: – Carry out the measurements for the standard solutions for a minimum of three concentration levels (close to the limit of quantitation, the middle range, and close to the upper limit of the linearity); – For each concentration, perform several repetitions (10 repetitions minimum is recommended*).
For reference material or test sample: – Perform several repetitive measurements (10 repetitions minimum is recommended*)

*The recommended number of repetitive values allows the use of sound statistical evaluation. However, one should be aware that there are situations when this is not possible due to the lack of a sufficient amount of sample or the high cost of a single measurement

Table 7.8 Typical values of precision depends on the concentration

Range of concentration	Precision expressed as relative standard deviation (%)	
	As reproducibility	As repeatability
≤1 μg/kg	≤35	≤55
<1 μg/kg ≤ 0.01 mg/kg	≤30	≤45
<0.01 mg/kg ≤ 0.1 mg/kg	≤20	≤30
<0.1 mg/kg ≤ 1 mg/kg	≤15	≤20
>1 mg/kg	≤10	≤15

considered as indicative, regarded as typical for a given range of concentrations. It is needless to state that in specific cases the relative error can diametrically differ from those indicated in the illustration.

7.16 Evaluation of the Accuracy and Precision

In order to evaluate the accuracy and precision of measurements for validated analytical methods, it is recommended that measurements for calibration solutions of

Table 7.9 Example calculations of the accuracy and precision for one standard solution of known concentration of 1.30 mg/L

Measurement's result, mg/L	Difference (Δ)	Δ^2
1.23	−0.07	0.0049
1.21	−0.09	0.0081
1.30	0.00	0.0000
1.59	0.29	0.0841
1.57	0.27	0.0729
1.21	−0.09	0.0081
1.53	0.23	0.0529
1.25	−0.05	0.0025
Mean = 1.36	Σ = 0.49	Σ = 0.2335

different concentrations and for test samples be performed, respectively. It is recommended that one of the standard solutions should be close but above the detection limit, whereas the second should be close but below the upper concentration within the linear range of calibration. Depending on the calibration dynamic range, those two standards will cover the whole range or the middle concentration should be also added. The test samples are used to examine the precision of measurements in the presence of the matrix. Due to the statistical requirements, it is recommended that no less than seven repetitions should be performed for each solution (Table 7.9).

The accuracy of the measurement (relative to the reference value of 1.30 mg/L) is calculated as 0.49/8 = 0.06 mg/L; and precision (as repeatability) is calculated as

$$\sqrt{\frac{0.2335}{(8-1)}} = \sqrt{0.03336} = 0.18\ \text{mg/L} \tag{7.1}$$

7.17 Evaluation of Recovery

Recovery testing is a very important part of documenting the reliability of measurement results. It is used for the evaluation of the systematic error and is used as a measure of the 'trueness' of the method. This can be described as a difference between the mean value of a number of repetitive measurements and the accepted reference value of the analyzed object. When it is expressed as a ratio of the mean value to the reference value (as one over one or as percentage), it is termed 'recover.' This can be measured either with the use of CRMs or given test samples. Evaluation of recovery can be done by comparing the average value of repetitive measurements for selected reference material, certified for the content of the analyte and closely matches the test sample with regards to matrix composition and content of analyte, with the certified value.

When well-matched CRMs are not available, it is recommended to perform spiking of a defined amount of analyte (most likely the pure chemical) to the sample. The measurements are performed for the non-spiked and spiked sample. From the difference of both quantity values, it is possible to estimate the calculated concentration of the added analyte. In the case of unbiased measurement procedure, the difference between both values would be equal to the defined amount of added analyte. When carrying out the recovery test, the following conditions should be maintained: analyte added to the sample should be present in the same chemical form in which it is present in the real sample, and the total concentration of the analyte after the addition of the standard should be within linear range of the calibration.

7.18 The Limit of Detection and Quantitation

The dynamic range of a calibration curve refers to the concentration/content of an analyte that can be determined using a given measurement procedure with acceptable accuracy and uncertainty. The upper concentration can be evaluated by examination of the course of linearity, as the lower concentration can be evaluated by examining the probability of distinguishing the analytical signal from the base line noise. Both values depend not only on measuring technique but also on the applied measurement procedure covering the entire analytical process.

In practice, it is convenient to define the scope for which both values are determined. When the primary characteristic of the given analytical technique is of interest, it is recommended to execute the measurements for blank as well as for matrix-free standard solution, which enables evaluation of the best instrumental performance of the technique. In the case the important information is the capabilities of the measurement procedure used for the matrix-rich test samples, then it is recommended to execute the measurements for the blank solutions that underwent the entire analytical process, as well for real test samples. With regard to the detection/quantification limits, typically the 'instrumental' values are lower than the 'procedural' ones.

7.19 The Limit of Detection

The limit of detection (LoD) indicates the smallest amount/concentration of the analyte that can be detected using a given measurement procedure. The LoD is often assumed to be threefold the value of the standard deviation for the blank sample or a sample containing a low concentration of the analyte. The numerical value of the LoD has the dimension of the concentration/content of the analyte.

There are several methods used for the calculation of LoD; it can be calculated as a sum of average signal for the blank and three times the standard deviation (blank + 3 s), or can be calculated from the slope of the calibration curve, so the detection limit is calculated as 3 s/b, where b is the slope of the graph (Table 7.10).

Table 7.10 How to evaluate detection limit: recommendation and examples

Kind of sample	How to proceed*
Blank for the matrix-free standards (only solvent and reagents used for the preparation of matrix free standards)	For 10 independently prepared blanks, perform 10 repetitive measurements Calculate the mean value and the standard deviation of the mean
Procedural blank (as above + all reagents used for the test samples preparation; execute the entire analytical process)	For 10 independently prepared procedural blanks, perform 10 repetitive measurements Calculate mean value and standard deviation of the mean
Test samples with low but above detection limit content of analyte	For 10 independently prepared test sub-samples, perform 10 repetitive measurements Calculate the mean value and standard deviation of the mean

*The recommended number of repetitive values allows the use of sound statistical evaluation. However, one should be aware that there are situations when this is not possible due to the lack of a sufficient amount of sample or the high cost of a single measurement

7.20 The Limit of Quantitation

Limit of quantitation (LoQ) indicates the lowest concentration/content of an analyte that can be determined with the accepted uncertainty, using a given measurement procedure. The LoQ is most commonly calculated as six or ten times the standard deviation for a blank sample or a sample containing a low but above LoQ concentration of the analyte. The numerical value of LoQ has the dimension of the concentration/content of the analyte.

There are several methods used for the calculation of LoQ; it can be calculated as a sum of average signal for the blank and six or ten standard deviation or can be calculated from the slope of the calibration curve, so the detection limit is calculated as six or ten of s/b, where b is the slope of the graph.

The limit of quantitation: perform measurements as described for limit of detection.

7.21 Robustness

One of the important features of validation is the robustness of the analytical procedure, meaning the stability of analytical results under the minor deviation in the procedure or environmental conditions. The robustness is usually evaluated for the finally accepted analytical conditions, where the possible range of random deviation can be predicted. Evaluation of robustness should indicate the most critical parameters that are the most vulnerable to changes influencing the final results.

Evaluation of robustness can be carried out in two ways:

1. By interlaboratory comparisons, with a sufficient number of participating laboratories ($\geq$10) using the same measurement procedure.
2. In the given laboratory, conducting a planned series of experiments, changing selected parameters of measurement procedure (including all personnel conduction given testing) and evaluating their influence on the analytical result.

The process based on the interlaboratory comparisons is expensive because it requires the involvement of many laboratories and resources needed for the data processing. Besides, for a number of testing it is not possible to find out the organized interlaboratory comparison or there are a lack of laboratories performing the given testing. Another way is to evaluate robustness inside of the laboratory, under the typical conditions (Table 7.11).

Evaluation of robustness proceeds as follows:

1. Identify the factors (variables) that are likely to have the greatest impact on the result;
2. For each factor, specify (predict) the maximum variation that can occur in routine work;
3. Describe a series of experiments under variable conditions;
4. Execute the experiment and assess the impact of each factor on the analytical result;
5. Optimize procedures (if necessary).

In this example, seven factors were indicated as those that could have a significant influence of final results. The systematic evaluation of robustness will enable in this case measurements to be taken under the conditions where those variables can be

Table 7.11 Example of the evaluation of the robustness of the analytical procedure

ID	Factor	Nominal value	Robust value
A	Mass of sample, g	5	10
B	Concentration of acid, mol/L	1	1.1
C	Time of stirring, min	10	12
D	pH	6.0	6.5
E	Temperature of reaction, °C	100	95
F	Time of reaction, min	5	10
G	Plasma gas flow in ICP-MS, L/min	28.5	30.0

Table 7.12 Scheme of the experiments enabling the evaluation of robustness of the analytical procedure

Value of parameters	Set of parameters							
	1	2	3	4	5	6	7	8
A/a	A	A	A	A	a	a	a	a
B/b	B	B	b	b	B	B	b	b
C/c	C	c	C	c	C	c	C	c
D/d	D	D	d	d	d	d	D	D
E/e	E	e	E	e	e	E	e	E
F/f	F	f	f	F	F	F	F	F
G/g	G	g	g	G	g	G	G	g
Result	s	t	u	v	w	x	y	z

identified. Let's assume that the nominal value of the factor is denoted by an uppercase character (e.g., A, B, …) and the robust value is denoted by a lowercase character (e.g., a, b, c, …). The typical manner for how to conduct the evaluation of robustness for eight identified factors is given in a Table 7.12.

The measurements should be conducted under eight sets of conditions with randomly but systematically selected nominal and random values of the critical factors. If set 1 was used, the result s was obtained, if set 2 was used, the result t was obtained, and so on. To evaluate the effect of a variable factor, one must select four results where the given factor value was nominal (uppercase) and the four results were the given factor was robust (lowercase) and compare the mean values for both sets of results. For example, to determine the impact of pH, the following values should be compared: the mean values for the results obtained under nominal conditions $(s+u+w+y)/4$ versus the mean value for the results obtained under robust conditions $(t+v+x+z)/4$. Such a calculation should be performed for all selected factors, then the values of the differences must be sorted out according to ascending or descending values, so as to note those with the most (significant) impact on the result. If there are no significant differences, calculate the mean and standard deviation of the eight results from s to z, which allows the robustness of the applied measurement procedure to be determined.

7.22 Summary

The validation of the measurement procedure is a tool used in a testing or calibration laboratory that allows the selected parameters of the given procedure to be determined, as well as confirming that they are suitable for its intended use.

Validation should cover sample preparation, measuring instrument, software, processing of the results and analyst competences. Analytical parameters of methods are calculated with the use of measurement results, and the resulting data are processed using appropriate statistical methodology. Validation should be conducted under conditions used for the testing of routine samples in the laboratory, most preferable with the use of samples deliver by clients.

Validation should be repeated in respect to the most critical parameters of the given measurement procedure, whenever something changes significantly in the conditions of the measurements (e.g., a new instrument, new software, a new type of objects, a new standard, new analyst).

The scope of the primarily validation and re-validation should be tailored so as to consider the intended purpose of the measurement procedure and the final use of the result, bearing in mind the cost and the risk of obtaining erroneous results.

Chapter 8
Measurement Uncertainty

Certain with uncertainty

In a chemical laboratory, tests are often conducted so that the final result is a mean of a number of individual measurements, and the experimentally determined standard deviation is used as a measure for the dispersion of the experimentally obtained data. Thus, the standard deviation determines the precision of the measurements since it shows how close the result was repeated in a given measurement series. It should be noted, however, that the precision of measurements does not indicate the accuracy of the result. It could happen that the mean value from even very precise measurements may differ significantly from the true value or the value recognized as the reference value.

In a chemical laboratory, tests are most often conducted in such a way that the given result is a mean value or a median of a set of values obtained within multiple measurements. The set of values from which the mean (or the median) is calculated might come from a series of repetitions for the single test sample or from a series of repetitions for a few test samples taken from the primal sample. As a measurement range, in which the values can occur, a standard deviation—determined experimentally—is used (variance or coefficient of variance are also used). Those values determine the precision of the measurements, as they show how well the result was repeated in a given measurement series. In Chap. 7, the terms 'accuracy,' 'precision' and 'trueness' have been discussed in detail. It is worth remembering that a high precision of measurements does not guarantee their accuracy.

The term of 'measurement error' is closely linked to performing measurements. According to the definition, 'measurement error' is the measured quantity value minus a reference quantity value (p. 2.16, VIM 3).

According to the basic axiom of metrology, there are no measurements completely free of errors. Therefore, when conducting measurements, one should be aware that the results are burdened with an error. Those errors contribute to the uncertainty connected to the course of the measurement process and its results.

E. Bulska, *Metrology in Chemistry*, Lecture Notes in Chemistry 101,
https://doi.org/10.1007/978-3-319-99206-8_8

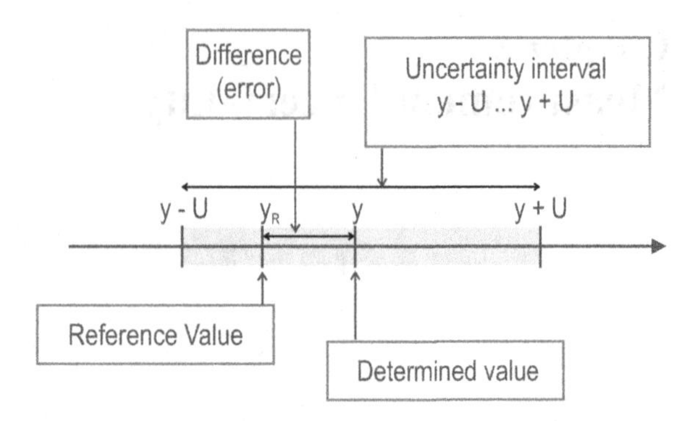

Fig. 8.1 Value, error, uncertainty

Generally, the error can be treated as a random value. The randomness of the measurement error makes it so that even the result of a measurement, conducted with the utmost care and to the best of knowledge, does not provide unambiguous information about the value of the measured quantity. In practice, the measurement error can be of a systematic and random character.

> **Measurement**: a process of experimentally obtaining one or more quantity values that can reasonably be attributed to a quantity.
>
> *clause 2.1; ISO/IEC Guide 99*

The result of a measurement is only an approximation (estimate) of the value of the measured quantity and, therefore, each result is accompanied by an uncertainty stemming from its randomness. It is believed that the measurement result cannot be expressed as only one number but, as it is used in the case of a random constant variable, should be expressed in the form of a range called the confidence interval. In such cases with a specified probability assigned to that range, it can be assumed that the value of the measured quantity is included within it—under the condition that all activities connected with the measurement have been done correctly of course. Such a range is also called the uncertainty range and is given in the form $(x - U; x + U)$, where x stands for the estimation of the value of the measured quantity—that is, its approximation obtained during the measurement, and U stands for the expanded uncertainty (Fig. 8.1).

The term uncertainty had been used for many years in measurements and it historically originates from the error theory and the error analysis. When calculating uncertainty, years of experience in physical measurements are used, where the term 'uncertainty' encompasses in its understanding all the elements influencing the result and occurring during measurement. The measurement uncertainty expresses the fact that for a given measured quantity and for the given result of the measurement of that quantity, there are infinitely many values distributed around that value, which is compliant with observations, data and knowledge of the laws of nature, and that can be assigned to the measured quantity at different levels of confidence.

Due to the complexity of the process of evaluating the measurement accuracy, different evaluation procedures were proposed by experts. In order to limit that diversity and to unify the process of the evaluation of measurement uncertainty, the International Bureau of Weights and Measures, with inspiration from the International Committee for Weights and Measures (CIPM) has undertaken actions that resulted in the publishing of a guide, ISO GUM *Guide to the Expression of Uncertainty in Measurements*, in which a uniform process of conduct was proposed when estimating the measurement uncertainty. An important document, in which various advice on the evaluation of uncertainty in analytical measurements can be found, is the guide issued by the Eurachem *Quantifying Uncertainty in Analytical Measurement*. Currently, the third issue of EURACHEM/CITAC Guide CG 4 (2012) is available.

The term 'measurement uncertainty' was accepted by international bodies and is currently commonly used for the description of the range of probability. It should be highlighted, however, that the term 'uncertainty' can have two meanings: colloquially, to express general doubts regarding measurement results; and in metrological meaning as a parameter determining the variability of the measurement results within a defined range. In the metrological meaning, the measurement uncertainty characterizes the justified spread of the value of the measured quantity and it should be determined for each measurement.

It is worth highlighting that the client—that is, the person interested in the measurement result—does not have to delve into the complexities of the evaluation of the analytical parameters of the measuring procedures used by the laboratory, but rather more into the process of evaluation of the quality of the results through the expression of the measurement uncertainty. Because of that, the efforts of the International Committee for Weights and Measures in the direction of establishing a uniform method of calculating and providing uncertainty are incredibly important and enable the comparison of results obtained in different laboratories.

It is worth mentioning here that in ISO 9000 standards, the following sentence concerning quality management systems in different contexts can be found: "measurement uncertainty is known and in accordance with the required measuring capacity.". Hence, everywhere that conducted measurements are subject to a measuring system, the required measurement capacity needs to be defined, and the obtained results must have an assigned measurement uncertainty. This especially applies to accredited laboratories.

8.1 Measurement and Its Uncertainty

The term 'measurement' encompasses a set of operations aimed at determining the value of a quantity. It means that there are a set of activities after which we can say that, at the moment of measurement conducted in specific conditions and after performing specified actions, the measured quantity X is within the specified range of values. This definition has an unquestionable axiom—that the value of the measurement has the form of a range on the dimensional axis. Thus, the issue is to determine the limits of that range. The term 'the correctness of the measurement' is understood as the compatibility between the mean value determined for a measurement series (Attention! In the VIM 3 dictionary it is referred to as "from an indefinite number of repeated values of the measured quantities") and the reference values. Hence, the question arises: how we can determine and/or classify the correctness of the measurements? According to the metrological principles, the correctness of the measurement is concluded on the basis of the uncertainty assigned to it.

The evaluation of the measurement correctness is one of the basic issues of metrology, as it defines the comparability, reliability and usefulness of the results.

According to the definition from the ISO GUM guide and the VIM 3 dictionary, uncertainty is understood as a non-negative parameter that characterizes the spread of the values that can, in a justifiable way, be assigned to the measured quantity. The measurement error, on the other hand, is the difference between the result of the measurement and the reference value of the measured quantity. The measurement error is a good indicator of the degree of compliance of the measurement result with the reference value, which is an indicator of the measurement correctness. The reference value is used for comparison with the measured values for a given quantity (object property). In chemical measurements, the reference value can be the certified value specified on the certificate of the certified reference material, the value from the certificate of the chemical standard of the pure substance, as well as the value obtained with the reference measurement procedure.

The absolute error ΔX: the difference between the measured value X and the reference value X_R; both are expressed in the unit of the measured value.

$$\Delta X = X - X_R \tag{8.1}$$

The relative error δX (expressed in percentage): the ratio of the absolute error ΔX to the reference value multiply by 100%

$$\Delta X = X - X_R / X \cdot 100\% \tag{8.2}$$

Let us assume that during the measurement of the X_p quantity, the result X was obtained. According to the axiom of the metrology, it can be noted that $X \neq X_p$. It is, therefore, necessary to complete the X result with the value of the parameter that characterizes the expected range of the results variability, connected to their random character. Only after the parameter is determined can we show the measurement result in the form $X_p = X \pm U$. Results given is such a form should also be completed with the probability assigned to that range, which determines that the true value of the measured quantity lies in the determined range.

In this context, two questions arise:

- How can the measurement uncertainty be determined?
- What uncertainty can be accepted, so that the measurement result can be useful?

8.2 The Reliability and Usefulness of the Measurement Results

It is clear that every measurement result should include an assigned uncertainty of measurement. A question arises, if a result includes an uncertainty, can it always be considered as **reliable and useful**?

A reliable result is one in the $[x - U; x + U]$ range in which the true value of the measured quantity is located. A result determined in such a way is a result that can be trusted. Measurements conducted in some laboratories can have a smaller uncertainty than those conducted in other laboratories. If, however, in two laboratories the uncertainties are determined according to the metrological principles, then both results will be reliable.

The reliable results require a detailed evaluation of all possible sources of the measurement uncertainty.

Useful results: is a reliable result always useful? In order to address this question, it is necessary to consider the goal of the measurement and the intended use of the results. Thus, the measurement result, with the assigned uncertainty, should be useful for a chosen purpose. Only reliable and useful results can be used when an important decision will be derived as to whether a given product was made properly, whether the production process runs properly, or to verify scientific hypotheses.

Useful measurement results shall therefore be accompanied with the well evaluated uncertainty, considered as the range where the result can still be useful for performing assessments or making decisions. With overestimated uncertainty, the risk of incorrect assessment can be high enough that, due to the costs of the wrong decision, the measurement result could be useless.

Definitional uncertainty: the component of measurement uncertainty resulting from the finite amount of detail in the definition of a measurand.

clause 2.27; ISO/IEC Guide 99

Target measurement uncertainty: measurement uncertainty specified as an upper limit and decided on the basis of the intended use of the measurement results.

clause 2.34; ISO/IEC Guide 99

Thus, in VIM 3, two terms related to the measurement uncertainty and regarding reliability and the usefulness of the measurement result are introduced. The term '***definitional uncertainty***' means, in practice, the minimum measurement uncertainty achievable in any measurement in a given measurand.

In literature, the terms 'definitional uncertainty' or 'basic uncertainty' can also be encountered. In practice, this means that taking into consideration all the steps of a given measuring procedure and including all components of the uncertainty, the result cannot be obtained with a smaller uncertainty than the definitional uncertainty. The term '***target measurement uncertainty,***' on the other hand, refers to the upper limit of the uncertainty that is acceptable for a specific use of the measurement result. In practice, this means that a useful result is one that has an assigned uncertainty that does not exceed the specified target uncertainty.

8.3 Measurement Results

By conducting a series of measurements in the conditions of repeatability, one obtains a set of raw data, the variability of which is a reflection of the random error. Admittedly, the error is a very useful term for understanding the rules and basics of the measurement uncertainty evaluation; however, its practical meaning is limited due to the lack of ability to determine the error for the given measurement. This stems from the fact that the true value of the measured quantity is unknown. In practice, the term 'reference value' is used; this is a value assigned to the determined quantity and recognized—sometimes arbitrarily—as the value determined with acceptable uncertainty for the specific use.

For the assessment of the numerical values obtained as the result of conducted measurements, it is accepted to use several terms and theorems of the theory of probability. In accordance with it, every measurement result is a random variable and the best model of a random variable is its probability spread. To describe the random variables, most often the expected value (μ) and the standard deviation (σ) are used. In practice, the values of those parameters are not known, and they are estimated on the basis of a series of experimental tests, and those are called estimators.

The expected value μ, which corresponds to the true value, is estimated by calculating its estimator from the results of the test, which is the mean value $X_{\text{mean.}}$

In mathematical notation, if each result is noted as $X_i, i = 1, 2, \ldots, n$; where n is the number of results, then the mean value is calculated on the basis of the dependency:

$$\overline{x} = \frac{\sum_{i=1}^{n} x_i}{n} \tag{8.3}$$

Standard deviation is the measure of the spread; and if we do not know the standard deviation for a population, we calculate its estimator s from the test results.

For large but finite set of results (populations of results), the standard deviation is calculated with an equation:

$$\sigma = \sqrt{\frac{\sum_{i=1}^{n} (x_i - \mu)^2}{n}} \tag{8.4}$$

where: n is the number of repetitions of the measured quantity, and x_i is the measurement value in ith repetition.

In this case, the true value lies within the range

$$\bar{x} - g\left(\frac{\sigma}{\sqrt{n}}\right) < \mu < \bar{x} + g\left(\frac{\sigma}{\sqrt{n}}\right) \tag{8.5}$$

where g is the constant describing the width of the range or the probability (p) of finding the true value within the range. Characteristic values of g: 1.00 ($p=0.683$), 1.96 ($p=0.950$), 3.09 ($p=0.998$).

It is known that if we want to estimate the trust range for any parameter of the population, then we should know the probability spread for its estimator. If the selected parameter, for which we want to estimate the trust range, is the expected value μ, then its estimator is the mean value x.

If the standard deviation is determined on the basis of a small number of measurement results, then a Student's t-distribution is used. That distribution is a function of only one parameter, called 'number of degrees of freedom' $\nu = n - 1$, where n means the number of the results. The function of the density of the t-distribution is very complex; therefore, in practice tables are used to determine the probability. With an increasing number of the degrees of freedom, the distribution becomes convergent with the normal distribution.

The distribution of the mean value is the Student's t-distribution with a standard deviation equal to the estimator of the standard deviation of the mean s. The Student's t-distribution becomes convergent with the normal distribution for $n \cdot \infty$ (in practice for $n \geq 30$).

The features of the Student's t-distribution

- Defines the probability of the occurrence of the x result in small measuring population;
- Maintains the position of the maximum of the normal distribution, but differs in height and width (depending on the number of degrees of freedom);
- True value lies within the range.

$$\bar{x} - t\left(\frac{s}{\sqrt{n}}\right) < \mu < \bar{x} + t\left(\frac{s}{\sqrt{n}}\right) \tag{8.6}$$

where s is the estimator of the standard deviation for small samples, expressed with the formula

$$s = \sqrt{\frac{\sum_{i=1}^{n} (\bar{x} - x)^2}{n - 1}} \tag{8.7}$$

t is the measure of the deviations of the distribution of a small group of measurements from the normal distribution, depending on the given probability (confidence level) and the number of the degrees of freedom (n − 1 for a series of repetitions).

8.4 Error Versus Measurement Uncertainty

Error and uncertainty are distinct concepts. They should not be confused.

The theory of measurement errors is based on quantities, such as the true value of the measured quantity and the measurement error. By contrast, the theory of the measurement uncertainty is based on experimentally determinable quantities—namely, the measurement result (which is the estimator of the value of the measured quantity) and the measurement uncertainty.

The measurement error is the measure of the difference between two specific values.
The measurement uncertainty is the measure of the spread of the measurement results.

The common tendency is to avoid the use of terms such as systematic error and random error. Undoubtedly, we can try to avoid them, and, for example, instead of analyzing the 'sources of errors,' we can analyze 'sources of uncertainty,' and instead of calculating 'systematic errors,' we can calculate the 'corrective factor.' That does not, however, influence the values of the calculated measurement uncertainties.

The essence of differentiation between the measurement error and uncertainty is that an error is a difference between two specific values, whereas the uncertainty is a parameter of the spread of measurement results. Therefore, the error for each measurement of the series, conducted in specified conditions, has a different value, whereas the uncertainty of given measurements is a constant, non-zero value, even if the error was a zero for one single measurement. That is why the two terms cannot be used interchangeably and both have their specific meaning. In the error theory, the equivalent of the uncertainty is the limit error of the measurement.

The measurement uncertainty determines the predictable limits of the variability of errors that could not be compensated for or eliminated.

8.5 Errors in Measurements of Chemical Quantities

A very important step of all the studies of the natural phenomena is the determination of the properties of a given substance (e.g., the chemical composition of water), mainly by performing measurements. In scientific research, observations are made through the measurements of the selected, characteristic quantity (e.g., the content of cadmium in water) and calculating its numerical value, expressed with the use of selected units (see Chap. 3). As practice shows, when performing the measurement of the same quantity multiple times, even with the utmost care, we always obtain a set of different numerical values. Those differences stem from the influence of various factors on the measuring process. It is, therefore, justified to query how the result of the measurement can be shown so that it reflects the true value to the best possible extent. It is also justified to query how confident we are of the obtained value. The science of measurement (metrology) defines the terms ***random***, ***systematic, gross error*** and ***measurement uncertainty***. In the natural sciences, error is not, however, a synonym of mistake. As highlighted previously, the measurement error is the difference between the measurement result and the true value of the measured quantity. The measurement error means an impossible to avoid uncertainty, inseparably connected with the essence of each measurement of physical (length, mass, time) or chemical (the content of the given component) quantities. The role of the analyst is to estimate its value and to make the uncertainty as small as possible.

When introducing the term 'error,' in accordance with the error theory, the true value of the measured quantity (the object property) is recalled. It is known that in practice we do not know the true value and we can only refer to a value that is the best possible approximation of the true value. Therefore, in the VIM 3 dictionary, the term 'the correctness of the measurement' was introduced, which refers to the reference value. On the other hand, the definition of the measurement error includes the comparison of the measured value and the reference value. In the comments on the definition (Clause 2.16, VIM 3), it is highlighted that the reference value can come from the measuring standard or be an agreed value of the true value. It follows that in practice we depart from using the term 'true value' in the definitive meaning, since the use of the term 'reference value' is more justified.

8.6 Types of Measurement Errors

The gross error is connected with the occurrence in the set of results of one that deviates significantly from the other values in the series. The sources of the gross error are most often significant and atypical disturbances of the measuring system or an error of the person conducting the measurements. If the difference between the individual result and the rest of the results in a given measurement series is significant, that result can be arbitrarily rejected or, in the case of doubt, appropriate statistical tests can be used to reject deviating results. (For the rejection of deviating results, statistical tests are used.)

The systematic error is connected, for example, with an incorrect setup of the measuring device, which means that the measured value is systematically underestimated or overestimated with a systematic influence of outside factors on the read (e.g., the influence of temperature on the pH measurement). A systematic error remains constant over a series of measurement and it cannot be reduced by increasing the number of replicate measurements.

Apart from gross and systematic errors, ***random errors*** are also important, meaning those errors that are the most responsible for randomly spread values when conducting multiple measurements of the same quantity (measuring series). All measurements, even those conducted with the utmost care, are subjected to the influence of various random factors. In that case, we distinguish two sources of measurement errors: outside factors (the influence of the surroundings, e.g., changes in temperature during measurements) and internal factors (e.g., the stability of the measuring device, the quality of the used measuring glassware). The effects connected with the influence of the outside factors can, to a certain degree, be controlled by the person conducting the measurements, whereas the effects connected with the influence of the internal factors are closely connected to the measurement itself and cannot be removed. Therefore, it is important to correctly calculate or estimate the uncertainty value, since that is the range in which we expect the true value to be.

At this point, it is worth emphasizing again the relationship between the term 'measurement error' and the term 'measurement uncertainty.' The error always defines the difference between the two values, and in the case of the measurement error, it is a difference between the true value and the value obtained as the result of the measurement. It follows that for each measurement result, the error will assume a different value (positive, negative or it can also be zero). Uncertainty, on the other hand, is a parameter that characterizes the spread of the measurement results—that is, one that determines the variability limits of the measurement results.

Measurement errors can be systematic and random
Systematic measurement error: a component of the measurement error that in replicate measurements remains constant or varies in a predictable manner.
Clause 2.17; ISO/IEC Guide 99

Random measurement error: a component of the measurement error that in replicate measurements varies in an unpredictable manner.

Clause 2.19; ISO/IEC Guide 99

8.7 Systematic Errors

Systematic errors can stem from:

- The work of the analysts whose experience and practice result in a specific procedure;
- Incorrect performance or adjustment of the measuring device (e.g., weight, measuring glassware, spectrometer);
- The feature of the analytic method (e.g., effectiveness of extraction, an incomplete run of the reaction).

Systematic errors cause a systematic bias of results and can cause their overestimation (positive errors) or underestimation (negative errors). Systematic errors, due to their nature (are constant in specified conditions) should be eliminated, if possible, from the measurement result. Of course, this is conditional upon the fact that their value can be determined.

In cases when it is possible to determine the value of the systematic error, for example, through comparison of the result obtained by a given method with the result obtained through the reference method, or through the analysis of reference material, it is possible to compensate for that value of the error in the measurement result. Such a procedure is valid when it is assumed that the systematic error has been determined correctly—that is that the assigned uncertainty is small, compared with the bias. That assumption would be justified if the systematic error was determined by using an infinite number of measuring results. Such a simplification can be applied in cases when it can be demonstrated that the uncertainty of determining the correction is very small in comparison to the observed spread of the results. In practice, the component connected with the correction dominates and is an essential component of the measurement uncertainty.

Systematic measurement error belongs to the category of influences that in metrological practice occur as correction or errors of the indications of the measuring devices. They are characterized by a value accompanied with associated uncertainty. In direct measurements, the measurement result is usually adjusted for the value of those systematic influences, and the measurement uncertainty includes only the random effects. Another procedure is also possible: namely, including the whole systematic effect into the confidence interval of the measurement result, and, therefore, treating it as a component of uncertainty. This is especially beneficial in indirect measurements, in which the appropriate adjustment for the correction value can change the definition of the measured quantity itself.

8.8 Random Errors

Random errors are a result of random variability of the value of the measured quantity. An important characteristic of random errors is that the positive and negative values of those errors are equally likely. Random errors occur as a result of many factors that fluctuate during subsequent measurements (e.g., temperature, pressure, voltage). The measure of the spread of the results due to the random errors is the standard deviation of the mean for a given measurement series. Standard deviation is one of the components—but not the only one—of the measurement uncertainty. Measurement uncertainty is a parameter of a broadly understood result spread as an effect of the influence of many partial random factors. In practice, it can happen that the resultant of those factors can be small in comparison, for example, to the resolution of the device. In such cases, the measurer will not observe any spread, which does not mean that the measurement uncertainty equals zero.

8.9 Requirements Concerning the Uncertainty

The knowledge of the values of the uncertainties assigned to the obtained results is essential for performing the comparison of the results between laboratories, clients and institutions that use the measurement results. Experienced laboratories can fairly judge their competences by the evaluation of uncertainties assigned to the provided measurement results. An acceptable value of the measurement uncertainty should always be assessed with a view of the specific requirements and should always be agreed with the clients (see: target uncertainty). It should also be borne in mind that in specific conditions even high values of uncertainty can be acceptable, and sometimes it is necessary to conduct measurements in a way that allows very low values to be obtained. Figure 8.2 demonstrates a comparison of results obtained in different kinds of laboratories (calibration, expert, testing) for testing the content of the given substance C [mg/kg] in a given matrix.

The best consistency of results was obtained for a group of calibration laboratories participating in key comparisons. It is worth noting that the measurement uncertainties provided by those laboratories are also closest to one other. The biggest dispersion of in the submitted results as well as their uncertainties can be observed for testing (named 'reserach laboratories' on Fig. 8.2) laboratories participating in inter-laboratory comparisons (ILCs); in between are the results from expert laboratories participating in proficiency testing (PT). This probably is a result of the fact that those laboratories that participate in PT handle the type of samples that are used in a given program, whereas those laboratories whose scope does not necessarily include such samples participate in ILC.

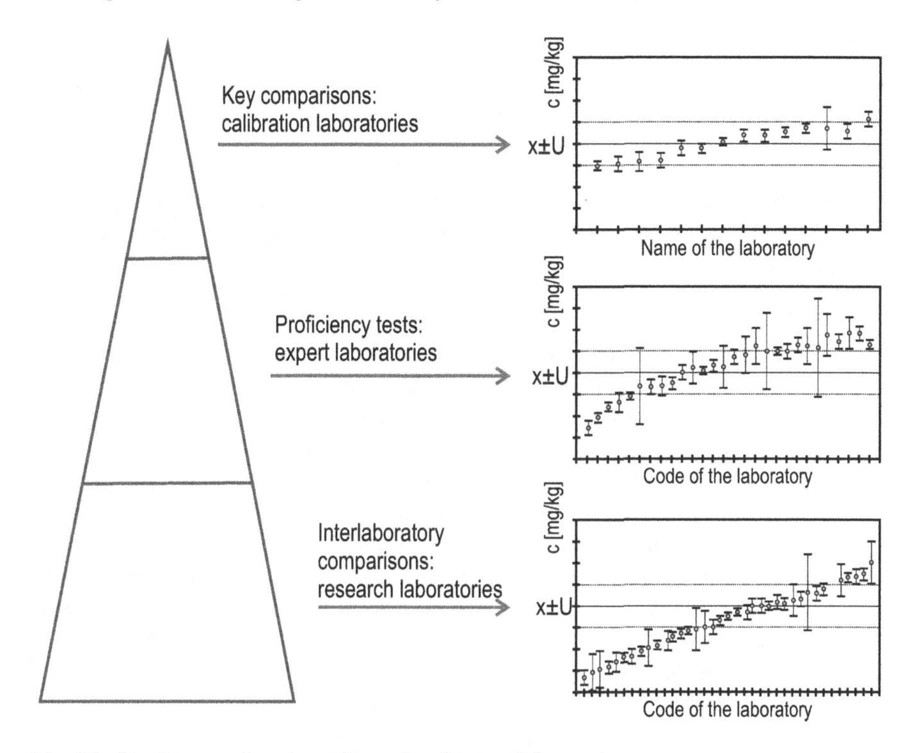

Fig. 8.2 Consistency of results within various kinds of laboratories

NOTE: In the key comparisons, organized by the International Bureau of Weights and Measures, participants are expected to submit their identification, which is publicly known. In proficiency testing or in inter-laboratory comparisons, the organizers ensure the confidentiality of identification of the laboratories.

The requirement to evaluate the uncertainty has significantly changed the approach of chemists to understanding of the quality of the results. Although it became common practice, one can still hear the opinion that the determination of uncertainty has been unnecessarily exaggerated. Some analysts still believe that the effort directed at the evaluation of uncertainty is unnecessary and that it is done because it is required in the standards and by accreditation. Often an opinion can be encountered that the determination of the uncertainty is very difficult and complicated: “We must do it, so we do it.” The requirement to provide the result with an uncertainty assigned to it is, in fact, very clearly articulated in the ISO/IEC 17025. It is worth noting, however, that the high expectation to determine the uncertainty enforces the necessity to conduct a detailed insight view into the used measuring procedure and critical evaluation of factors influencing the quality of the result. Submitting the result with the assigned uncertainty of the measurement is the confirmation of its reliability, hence the effort involved in the evaluation of the measurement uncertainty is fully justified substantively.

The result of the measurement can be considered reliable only when it is provided with an assigned uncertainty determined in accordance with the procedure described in the ISO GUM guide.

As previously mentioned, the evaluation of the measurement uncertainty is a rational process of evaluating its reliability. Hence an increasing number of laboratories conducting chemical measurements pay attention to that issue. The process of evaluation of the uncertainty requires the analyst to critically estimate all steps of the measuring procedure that can be the source of the uncertainty and can have a significant impact on the measurement result. For the end-users of the results, the value of the uncertainty shows the range in which the result can be reliable. It should be clearly emphasized that a correctly conducted evaluation of the uncertainty allows the measuring procedure used to be critically assessed. Providing a result with assigned uncertainty is extremely important when results are compared, especially when decisions made on that basis relate to levels in the area close to the limiting values (e.g., the highest acceptable concentration of a given substance).

To summarize the above considerations, it can be said that the ability to correctly calculate the result uncertainty becomes an indispensable tool in the everyday work of the analyst. The correctly evaluated values of the uncertainty are also indispensable in determining the cohesion of the measurements. As previously mentioned, the commonly accepted procedure for calculating uncertainty is the one described in the GUM guide. Issued by the ISO, the *Guide to the Expression of Uncertainty in Measurement* (ISO GUM) defines the generally accepted process of calculating and expressing uncertainty in different types of measurements. According to the described guidelines, the procedure includes the identification of potential factors that influence the measurement result, the determination (on the basis of own experimental data or on the basis of literature) of their numerical values, the consideration of the source of uncertainty and calculation of the value of the combined standard uncertainty and then of the expanded uncertainty. The procedure uses appropriate mathematical (statistical) tools that enables the inclusion of various factors that influence the result. Thanks to that, it is possible to obtain a numerical value (uncertainty) including all previous measurement results, current experimental data, information from the manufacturer of the equipment or standards and literature data. Remember that providing the result with an assigned uncertainty does not mean that the value of the uncertainty must be small.

It is worth quoting a statement here from the ISO GUM guide § 3.4.8:

"the calculation of uncertainty is not either a routine task or a strictly mathematical task. It depends on the detailed knowledge of the nature of the measured quantity and the measurement procedure..."

In chemical measurements, a significant source of the uncertainty can be the way that samples are prepared or the matrix effects influence the result of the determination (interferences).

When does a need arise to determine the uncertainty again? It is obvious that it does not concern every determination, but uncertainty should be established in the following scenarios:

- When the laboratory introduces a new measuring procedure;
- When the measurement conditions change (new device, new employee);
- During the validation of the measuring procedure.

The resulting uncertainty value, determined for a given measuring procedure, can then be used when the results are provided to the clients. Hence, it is worth emphasizing once more that it is not necessary to determine the uncertainty of every single measurement; once determined, uncertainty can be used for all results obtained with the use of a specific measuring procedure in specific conditions.

8.10 Determination of the Measurement Uncertainty

In many fields, not only in science, many decisions are made on the basis of analytical results. It is thus obvious that more and more attention is paid to their quality. The uncertainty is the property of each measurement results, is associated always with measurements and is coming out on each steps of measuring procedure. It is, therefore, not a property that is supposed to add additional difficulty to the measuring process. The measurement uncertainty results from the uncertainties of all activities performed during the analytical procedure. By evaluating the combined uncertainty, it is therefore necessary to examine the sources and types of uncertainties for individual steps of the analytical procedure.

The uncertainty of the measurement result demonstrates its reliability and is used for the comparison of results provided by different laboratories and/or carried out at different times. It is thus important to conduct the determination of the uncertainty in the same way. That is why the aforementioned ISO GUM guide was issued, as it describes the guidelines regarding the definition, assessment and recording of the uncertainty of the measurement result. The requirements of the ISO standards in the scope of determination and expression of the resulting uncertainty are obligatory for accredited laboratories. The unified—on the international forum—convention regarding the theory of uncertainty (ISO GUM) has been commonly accepted, not because it is an imposed law, but most of all, because of the advantages of the solutions and the benefits from its broad use. The process of evaluating of the uncertainty of the measurement results, described in the ISO GUM guide, is a universal method, possibly to be used regardless of the field of science or technology.

Uncertainty encompasses many factors that influence the result. Some components of the uncertainty can be obtained by statistical analysis of the data from repeated measurements and can be expressed as the standard deviation; others are estimated on the basis of the probability distribution. The most often used distributions are the normal (Gaussian), triangular or rectangular distribution.

Providing the result with the assigned uncertainty is required in the ISO/IEC 17025. But it is not the only argument in favor of using uncertainty. The procedure of evaluating the uncertainty includes a very detailed analysis of the measurement procedure, which is connected to the necessity to select and assess various factors that influence the final result. That aspect of uncertainty forces the analyst to consider the influence of various factors, and that can, in turn, be a good basis for the selection of the work strategy in the case of a need to improve the analytical parameters of the method. Therefore, during the analysis of particular parameters, it is possible to assess which of the included factors is critical. This allows appropriate actions to be taken that modify the measuring procedure used.

Evaluation of the uncertainty of the measurement result requires, most of all, a good knowledge of the applied measuring procedure and the infrastructure in a given laboratory. An essential part of the evaluation of the uncertainty is the selection of all the parameters that influence the final result. Among the most often included parameters is the quality of the reagents or the reference materials and the performance of the measuring devices used.

Commonly identified sources of uncertainty of chemical measurements are as follows:

- Incorrectly defined measurand;
- Sample does not fully represent the feature of the entire object;
- The measurement procedure is not used correctly;
- Systematic errors occur;
- Lack of knowledge of the influence of environmental conditions and their fluctuation on the result of the measurement;
- Uncertainty of the calibration of the measuring device;
- Resolution of the measuring device;
- Uncertainty of the certified value of the reference standards and/or the reference materials;
- Uncertainty of the physical constants and/or atomic masses used in the calculations of the final result.

The measurand is a quantity intended to be measured and it depends on the analyzed object and its feature to be tested. The definition of measurand for a given test is not always trivial; definitely it should be related to the aim of performing testing. In respect of the test result, the measurement uncertainty should always be given since it is considered to be an intrinsic part of the measurement result. The numerical value of the combined uncertainty allows an independent and objective interpretation of the results of testing. It can be also used for the metrological comparison of results and, last but not least, to evaluate their quality and usefulness for the intended use.

For each of the parameters, their influence is on the analytical result should be assessed and the value of the standard uncertainty should be calculated. A detailed uncertainty budget can be therefore used for the optimization of the measurement procedure aiming mainly for an analytical performance. The determination of the combined standard uncertainty is based on the law of propagation, which means the sum of the variance—that is, the square root of the values of the standard uncertainty. The breakthrough in the introduction of the uncertainty was the recognition that when summing the variances, both the results obtained in the laboratory and the manufacturer's data should be included; for example, the purity of the reagents, the uncertainty of the certified value or the previous experiences of the laboratory. In relation to that, two methods of evaluating the uncertainty are distinguished: type A and type B. The type A evaluation includes the statistical analysis of measured results obtained under defined conditions; the components of the type A uncertainty are expressed as a standard deviation with a known number of degrees of freedom. The B type evaluation includes other data, determined by means other than statistical analysis. All the values belonging to the B type, before using the propagation law, need to be reduced to the standard conditions—that is, to include statistical coefficients and predicted distributions (normal, triangular, rectangular).

Type A evaluation of measurement uncertainty: evaluation of a component of measurement uncertainty by statistical means.
Type B evaluation of measurement uncertainty: evaluation of a component of measurement uncertainty by another means.

In practice, it means that the type A uncertainty has an experimental character and it can be used when results of a series of measurements conducted in a laboratory with the help of a defined measuring procedure and in specified conditions are available. The type B is of the calculation type and can be used when access to credible information about the value of the uncertainty or another means of description of the spread of the values for a given quantity is available. Type B can be used for the following information; for example, authorities published quantity values, obtained from a calibration certificate, obtained from the accuracy class of a verified measuring instrument.

In order to normalize all values to the standard deviation of type B, in practice, the rectangular and triangular distributions are used. For example, the concentration of a component of the solution provided by the manufacturer is 1000 mg/L $\pm$ 2 mg/L, which means that the value can be found in the range 998–1002 mg/L. Taking into consideration that, in such a case, the probability of occurrence of a result is equal in the whole range, we take the rectangular distribution; hence the standard uncertainty is $2/\sqrt{3}$—that is, 1.16 mg/L.

In the case where we can assume that the probability of the occurrence of a value close to the mean value is bigger, it is recommended to use the triangular distribution. For example, the volume of the measuring flask provided by the manufacturer is 100 mL ± 0.1 mL. The nominal value is the most probable; hence, taking the triangular distribution, the standard uncertainty is $0.1/\sqrt{6} = 0.04$ mL. After the normalization of the statistical parameters and calculating the absolute values into the relative values, we can sum the variances—that is, the square roots of relevant values of the standard uncertainties. After evaluating the square root of the value, we obtain a combined standard uncertainty for a given measuring procedure (in specified conditions). Knowing the value of the combined standard uncertainty, we can, through correlation with strictly statistical parameters, asses that it is a range in which the true result lies with the probability of approximately 68%.

That value is very useful when there is a need to compare results on the basis of their standard uncertainty. However, when submitting the results to its end-users, it is recommended to submit the expanded uncertainty—that is, the value of the standard uncertainty multiplied by the coverage factor k, determined for the specific level of trust. In the laboratory practice, the most often used is $k = 2$; $k = 3$ is used less often. The determined value of the expanded uncertainty does not have a direct bearing on the degrees of freedom, as in that case we do not have the information about the number of repetitions. Nonetheless, it is assumed that, in approximation, the use of the factor $k = 2$ or $k = 3$ allows the range of uncertainty that corresponds to the probability of finding the result in specified range of 95% or 99% respectively, to be determined.

Uncertainty is a parameter that includes, apart from the precision of the measurement, many other factors that influence the changeability of the measurement result.

Most of the measured quantities in the field of chemistry are determined indirectly through the measurements of other quantities measured directly. The measured quantity, designated with the symbol y, is called the output quantity, and the quantities x_i (for $i = 1, 2, \ldots, N$) are the input quantities. The output quantity and the input quantities are treated as random variables, for which the probability spread needs to be evaluated. Moreover, two other parameters should be determined, namely the expected value and the standard deviation. For each input quantity x_i, the following values are of importance: the mean value x_{mean} and quantities influencing the input quantity. The influencing quantities are characterized with zero expected values and always non-zero standard deviation. Among influencing quantities, we can include the result spread or the bias of the measuring procedure, caused, for example, by a systematic error of the measuring device.

The measure of the spread is usually the standard deviation of the experimentally obtained results; therefore, normal distribution is assigned to it. In each case, when a sufficiently large set of results is available, it is possible to use the statistical assessment of the spread. When information on the resolution is given, the rectangular distribution is used, assuming that any value from the range is equally probable. This is also assumed in the case of correction factors and errors. Knowing the value and sign (plus or minus), those influences can be treated as random, with assigned the rectangular distribution with the full width at half maximum (FWHM), equal to the limiting values of those influences.

Expanded uncertainty is calculated by multiplication of the coverage factor and the combined standard uncertainty:

$$U = k \cdot u_c(y) \tag{8.8}$$

Expanded uncertainty is given for an arbitrarily defined confident level. In most cases, it is accepted as 95%. The value of the coverage factor for a given confidence level is determined by the probability distribution of the output value.

The combined standard uncertainty can be composed from a number of constituent uncertainties. Some of them can be determined on the basis of the results of a measurement series, characterized by a spread. Other constituents of the combined standard uncertainty, which cannot be assessed on the basis of the obtained spread of results—for example, uncertainties stemming from imperfections of the measuring equipment—are also evaluated by standard deviations, calculated on the basis of predicted probability distributions.

Those two groups of uncertainty, different in the way they are obtained, are a criterion according to which the uncertainties are divided into type A, which are determined with the help of statistical methods, and type B, which are determined through the use of other methods.

A commonly used way of evaluating uncertainty is to describe the measuring procedure in the form of a mathematical equation (model equation), which includes the input quantities (factors that influence the result) and output quantity (the measured quantity). A mathematical model of measurement is expressed with the functional dependency

$$Y = f(X) \tag{8.9}$$

where Y is a single output quantity, and X represents N input quantities. Each input quantity Xi (from X_I to X_N) is a random variable with the expected value of x_i.

Usually, the symbols of quantities are designated with capital letters, X and Y, respectively, and their estimates with small letters, x and y, respectively.

8.11 Evaluation of Uncertainty: Requirements in Chemical Measurements

The procedure of determination of measurement uncertainty, described in the ISO GUM guide is commonly recognized. The ISO GUM procedure is often called as 'modeling,' as the main requirement is the establishing of a mathematical equation (model) that describes the measuring process. In practice, the mathematical model is an equation that is used to calculate the measurement result. It is assumed that all constituents of that equation (input quantities) are parameters that influence the value of the measurement result, and thus they influence the measurement uncertainty. Therefore, for each input quantity, their values need to be assessed and the estimates of their uncertainties need to be provided. After normalizing all uncertainties to their standard values, the combined standard uncertainty should be calculated, using the law of propagation. The final step is the calculation of the expanded uncertainty for the specific confidence level.

The procedure of evaluation of measurement uncertainty, described in the ISO GUM guide, does not require knowledge of very advanced statistical methods. In reality, only basic information is necessary to conduct all the required calculations. The most important ones include knowledge of the propagation law and the ability to use different statistical distributions. When calculating the constituent values to standard uncertainties, it is necessary to be able to choose the type of distribution that a given value is subject to. It must be remembered that the type of distribution (normal, rectangular, triangular) influences the transformation of a given value into the standard form. This particularly applies to the type B uncertainty (e.g., archival measurement data, manufacturer data, literature data).

Typically, the process of uncertainty evaluation includes following steps:

- Specify the measurand;
- Specify the measurement procedure and describe the measurement model (equation);
- Identify the sources of uncertainty;
- Assign the numerical values to the uncertainty components;
- Calculate the combined standard uncertainty;
- Calculate the expanded uncertainty (with a given coverage factor, k);
- Examine the uncertainty budget.

This is how one can outline the basic scenario for determining the uncertainty in accordance with the procedure described in the ISO GUM guide. Since this document has been published, discussions have been conducted among the chemist community on the possibility of using the mathematical model to describe all steps of the measurements procedure. In chemical measurements, it is not always possible to describe the entire procedure in the form of a mathematical model and hence great effort was taken to develop a non-mathematical way to evaluate uncertainty (Table 8.1). In practice, a good alternative are those processes that use experimental data; often those are

Table 8.1 Evaluation of uncertainty of the measurement result

Process used for evaluation	Evaluation of uncertainty in practice	
Mathematically	Modeling	Requires a model equation, which includes all factors influencing the measurement result (ISO GUM)
Experimentally	Single-laboratory validation	Allows the use of all data collected during the validation of the measuring procedure, mostly the intra-laboratory repeatability and reproducibility, the bias of the method and/or the recovery
	Inter-laboratory validation (interlaboratory comparisons)	Allows assessment of the uncertainty on the basis of inter-laboratory reproducibility
	Laboratory proficiency testing	

Table 8.2 General description of the procedure used for the evaluation of uncertainty in practice

Mode of evaluation	Comments
Modeling	The uncertainty of a single result obtained in a given laboratory (does not include the fluctuation of the conditions over time)
Single-laboratory validation	The uncertainty that characterizes the measuring procedure in a given laboratory (includes the fluctuation of the conditions in a given laboratory over time)
Inter-laboratory validation	The uncertainty that characterizes the spread of results obtained in different laboratories (high variability of measuring conditions—for different laboratories)
Laboratory proficiency testing	

data collected by the laboratory in the process of validation, during quality control and through participation in inter-laboratory comparisons.

In the following section, the advantageous and limitations of the most often used methods for the evaluation of the uncertainty in chemical laboratories will be given. It should be noted that, depending on the method used, uncertainty has a different meaning and either refers to a single result or represents a spread of results (Table 8.2).

Evaluation of the uncertainty, regardless of the procedure used, requires full engagement, good knowledge of the measuring procedure, advanced knowledge on the tested object (the sample) and the knowledge of the respective statistical procedure.

8.12 Modeling According to the ISO GUM Guide

The process of evaluation of uncertainty, according to ISO GUM guide ('Guide to the Expression of Uncertainty in Measurement') includes the following steps:

1. Determination of the measuring procedure and the measured quantity.
   The quantity of interest in a given measurement should be clearly defined, and the unit for the expression of final results should be selected.
2. Description of the measuring procedure in the form of a mathematical equation (model).
   The model equation is a mathematical description of the dependency of the value of the determination result and the measured values. That dependency has the form of the equation:

$$y = f(x_1, x_2, \ldots X_n) \quad (8.10)$$

   where y is output quantity; x_1, x_2, …, x_n are input quantities.
3. Identification of factors that influence the measurement result (uncertainty).
   All identified sources of uncertainty—those that influence the measurement—should cover all factors that affect measurement results. Those could include, for example, the recovery of the analyte from the sample, the conditions of sample storage, the purity of the used reagents, stoichiometry of the reactions, the conditions in which the measurements are conducted, the precision of the measurements, the stability of the measuring device, the resolution of the measuring device, the quality of the standards used (i.e. the uncertainty of the certified value).
4. Assigning the source of uncertainty to type A or B (for the purpose of normalization to standard uncertainty).
   Uncertainty components can be divided into two categories, depending on the method of calculating their numerical values.

Type A—can be calculated with the use of statistical methods (they are characterized by estimates of the standard deviation and the number of the degrees of freedom); these are the quantities whose values and uncertainties can be determined on the basis of the results of measurements conducted in the laboratory.

Type B—have been determined not by statistical means (they are characterized by the approximate estimates of the standard deviation); these are the quantities whose values and uncertainties have been entered into the measurement from sources other than statistical ones (e.g., calibration certificates, certificates of the reference materials, tables).

This classification is not connected with the different nature and/or properties of the components of uncertainty, nor with their indication; it is only a way to determine the way to normalize all values into the standard uncertainty. For uncertainty of the type A, the estimate of the standard deviation is calculated from the repeated measurements; for the type B, the available data, other than the measuring ones are used.

8.13 Evaluation of Standard Uncertainty for Each Identified Component of the Measurement Uncertainty

Each of the quantities is characterized by a name, unit, value, standard uncertainty and the number of the degrees of freedom. In principle, two methods are used to calculate the standard uncertainty, In the case of using the type A method, the value of the standard uncertainty equals the standard deviation of the arithmetical mean. When using type B to estimate the uncertainty, its value is strictly connected to the probability distribution that describes the spread of the variable.

8.14 Calculating the Type A Uncertainty

The quantity X, measured directly, is treated as a random variable. Conducting direct measurement is the equivalent of drawing n-element sample $\{x_1, x_2, \ldots, x_n\}$ from an infinite population, which comprises all possible measurements. We assume, by default, that the general population has the normal distribution $N(\mu, \sigma)$, where μ is the expected value, σ—standard deviation. For the result of the measurement, the numerical value of the estimate of the expected value is taken, which in practice means the arithmetical mean of the measurement results.

$$\bar{x} = \frac{1}{n}\sum_{i=1}^{n} x_i \tag{8.11}$$

The standard uncertainty of the result of the measurement (expressed as a mean value) of the x quantity is called the experimental standard deviation of the arithmetical mean, which is calculated by the equation

$$u(x) = s = \sqrt{\frac{1}{n(n-1)}\sum\nolimits_{i=1}^{n}(x_i - \bar{x})^2} \tag{8.12}$$

The uncertainty calculated this way is the standard uncertainty calculated with the type A method.

8.15 Calculating Type B Uncertainty

The standard uncertainty is estimated as type B in the case where only one measurement result is available, or when the results do not demonstrate any spread. Then the standard uncertainty is estimated on the basis of the knowledge about the quantity or the range in which the true value should be. In the case of results not demonstrating any spread, the main basis for the measurement uncertainty is the calibration uncertainty $\Delta_d x$ equal to the value of scale interval of the measuring device used. It is assumed that the $\Delta_d x$ is equal to half of the width of the rectangular distribution and the standard deviation is

$$u(x) = \frac{\Delta_d x}{\sqrt{3}} \tag{8.13}$$

(an estimate of the standard deviation in the uniform distribution). If on the basis of general knowledge, the symmetrical triangular distribution can be taken, then

$$u(x) = \frac{\Delta_d x}{\sqrt{6}} \tag{8.14}$$

The other reason for the measurement uncertainties not showing any spread is the uncertainty of the experimenter, owing to causes beyond their control. The experimenter is using their own experience and knowledge in order to determine the uncertainty and the standard uncertainty stemming from it. Often the experimenter's standard uncertainty is also estimated on the basis of the rectangular distribution; in this case:

$$\mathrm{u}(x) = \frac{\Delta_e x}{\sqrt{3}} \tag{8.15}$$

The data taken from the literature, mathematical tables or values calculated with the help of a calculator are also burdened with uncertainty. If the value of the experimental standard deviation is not given (if it is given, then the uncertainty $u(x)$ is equal to that deviation) and there is a lack of any information on the uncertainty, standard uncertainty is then calculated from the formula, using the rectangular distribution:

$$u(x) = \frac{\Delta_t x}{\sqrt{3}} \tag{8.16}$$

When the values of results are characterized by the uniform distribution (rectangular), then the value of single result is assumed to be in the range $-a \ldots +a$, with an equal probability; in that case, the value of the standard uncertainty is: $a/\sqrt{3}$ (where a is half the width of the range $-a \ldots +a$).

When the values of results are characterized by the triangular distribution (the value of single result is in the range $-a \ldots +a$, but the occurrence of the mean value from the range is the most probable), an, in this case, the value of the standard uncertainty is a $a/\sqrt{6}$.

Statistical distributions used for the evaluation of standard uncertainty
Normal distribution (known as *Gaussian distribution*): is used when a large set of data are available, which depends only on the randomly distributed parameters. In normal distribution, most of results are clustered symmetrically on both sides of a central value, as fewer occurred unbounded on both sides. Values near the mean are more likely than values far from the mean. Standard uncertainty is equal to the standard deviation, with a confidence level of 68.3%
Rectangular distribution: all values lie with equal probability in the given range between $-a$ and $+a$. Thus, estimated standard uncertainty can be calculated as $u(x) = a/\sqrt{3}$. Rectangular distribution should be applied for the information given in certificates or other documents, where the information that the value lies between the range is given. e.g., the purity of cupper standards is quoted as (99.99 ± 0.01)%
Triangle distribution: this describes the situation where it is expected that values near the mean are more likely than those far from the mean, close to the extremes of the range. Thus, the estimated standard uncertainty can be calculated as $u(x) = a/\sqrt{6}$. Triangle distribution approach should be applied for the data given in the specification of e.g. volumetric glassware volume is quoted as (100 ± 0.1) mL (in 20 °C).

8.16 Calculating Combined Standard Uncertainty

In cases where the measurement result is calculated from values of other quantities, the law of propagation of standard uncertainties of all components is applied, which results in combined standard uncertainty. Assuming that the input quantities are independent of one other (uncorrelated), the combined standard uncertainty is calculated by the use of the following equations:

$$u_c(y) = \sqrt{\sum_{k=1}^{K} \left(\frac{\partial f}{\partial x_k}\right)^2 \cdot u^2(x_k)} \quad (8.17)$$

$$u_c^2(y) = \sum \left(\frac{\partial f}{\partial x_i}\right)^2 \cdot \left(u^2(x_i)\right)^2 \quad (8.18)$$

If both methods of evaluation of uncertainty—type A and B—were used, then the following equation should be used to determine the combined standard uncertainty:

$$u(x) = \sqrt{u_A^2(x) + u_B^2(x)} = \sqrt{\frac{1}{n(n-1)} \sum_{i=1}^{n} (x_i - \bar{x})^2 + \frac{(\Delta_e x)^2}{6} + \frac{(\Delta_e x)^2}{3}} \quad (8.19)$$

8.17 Calculating the Combined Uncertainty (for the Selected Coverage Factor *k*)

The concept of expanded uncertainty, noted as ***U*** has been introduced to determine the range that encompasses a sufficient probability of the spread of values that can be assigned to the measured quantity in a justified way. The expanded uncertainty ***U*** is obtained through the multiplication of the value of the combined standard uncertainty $u_c(y)$ by the coverage factor ***k***.

The coverage factor ***k*** can have different values, depending on the required confidence (probability) level. The ***k*** value in the range of 2–3 is the one used most often, which with the assumption of a normal distribution, means a trust range of approximately 95% or 99%, respectively. The measurement result is provided as

$$Y = y \pm U \tag{8.20}$$

which means that the best approximation of the measured quantity ***Y*** is ***y*** and that a majority of the value spread of the measured quantity is in the range from $y - U$ to $y + U$.

Selected terms used for expanded uncertainty are listed below.

Expanded measurement uncertainty: the product of a combined standard measurement uncertainty and a factor larger than the number one.*

Clause 2.35; ISO/IEC Guide 99

Coverage interval: the interval containing the set of true quantity values of a measurand with a stated probability, based on the information available.

Clause 2.36; ISO/IEC Guide 99

Coverage probability: the probability that the set of true quantity values of a measurand is contained within a specified coverage interval.

Clause 2.37; ISO/IEC Guide 99

Coverage factor: the number larger than one by which a combined standard measurement uncertainty is multiplied to obtain an expanded measurement uncertainty.**

Clause 2.38; ISO/IEC Guide 99

* The term 'factor' in this definition refers to a coverage factor.

** Coverage factor is usually symbolized by *k*

8.18 Reporting the Results

When all measurements are completed and laboratory collected all set of the data refereeing to the conducted work, those should be presented in a form of final report. The results of measurements for a given sample or set of samples should be described so as to include all the information relevant to the end users of those results.

The content of the report, whenever a name is given for such a document, depends on the purpose and the end-user of the results. In the case of the internal use of results, the report can be simplified, meaning that it contains only identification of samples and the test or calibration results, with assigned uncertainty whenever relevant. In many cases, the description of the measuring procedure should also be added.

However, in a number of legal sectors, other formal information is also required; for example, the name and address of the laboratory conducting the tests, the name and address of the customer, information about the analytical procedure(s), tests results with units and, whenever applicable, a statement on the uncertainty of the measurements. Sometimes the date when the tests were performed and the environmental conditions can be relevant.

8.19 Evaluation of Uncertainty by Mathematical Model

In the process of measurements, the uncertainty is due to the occurrence of random and systematic effects. In order to estimate the overall uncertainty, it is necessary to identify all possible sources of uncertainty, calculate their estimates and, finally, combine them so as to obtain a value representing all effects. The procedure for the determination of the uncertainty, which is based on the modeling of the measuring procedure, has many advantages. The modeling means to construct the mathematical equation used to calculate the final result of measurement. In the best cases, the mathematical equation should cover all steps of the analytical procedure, reflecting all factors that influence the uncertainty of the measurement result. Hopefully, it should also be possible to collect relevant information on the values of the quantities influencing the uncertainty.

In measurements of chemical quantities, one can define also some other parameters that are not included directly in the equation, but still influence the uncertainty, despite the fact that it is possible to include them in the model equation.

In practice, the model equation often does not include quantity that reflects, for example:

- Heterogeneity of the sample;
- Variability of the composition of the matrix in the series of samples;
- Storage conditions;
- Influence of the matrix components (interferences);
- Memory effects.

Example 1 The uncertainty of weighing is determined through the propagation of uncertainties of all constituents that influence the measurement of mass. Most often, among the factors that influence the uncertainty, one can list uncertainty of the balance (calibration certificate of the balance), uncertainty of the weight used for intermediate checking of the balance (calibration certificate of the weight); and uncertainty of weighing in the laboratory (standard deviation of the control chart used for the long-term checking of the balance).

In the case of the analytical balance, with the accuracy of 0.0001 g, the expected value of uncertainty can be assumed to be in the range 0.0003–0.0005 g.

It should, however, be highlighted that an uncertainty reflects the weight of an 'ideal' object—that is, the weight. In reality, objects weighted are not 'ideal' which can significantly influence the weighing uncertainty.

Factors influencing the uncertainty when weighing real objects:

- Weighing of hygroscopicobject (sample);
- Weighing of object easy to collect electric charge;
- Weighing of object containing volatile components.

Table 8.3 shows the results of weighing two objects: a weight with the mass of 1.000 g and a soil sample (a sample of soil with the mass of 1 g has been placed in the vessel). In the case of the measurement of the mass of the weight, the result is the value read directly. In the case of the measurement of the mass of the soil, the result is the value after the tare is subtracted (the mass of the empty vessel). For each object, 12 subsequent weighings were performed.

BE AWARE! When measuring real objects (e.g., soil), the uncertainty (expressed as a standard deviation) can come to a value much higher than the uncertainty obtained during calibration with standard weights (as provided in the certificate of balance).

Example 2 The uncertainty of measuring a specified volume of liquid with measuring glassware is commonly taken from the calibration certificate of the pipette or from the laboratory results of the measurement of mass of the pipetted liquid.

For example, for a pipette of the type A class, of 25.00 mL, the uncertainty taken from the manufacture calibration certificate, is 0.03 mL. It should, however, be mentioned, that the calibration is executed with an 'ideal' object—that is, the distilled water with a specified temperature. In laboratory practice, pipetted liquids are not 'ideal,' which can significantly influence the uncertainty of the measurement result.

Table 8.3 Results of weighing various objects (standard weight and soil sample)

No	Weight of 1.000 g		No	Soil sample, of ≅1 g	
	Mass, g	Δ*		Mass, g	Δ*
1	1.0001	0.0003	1	0.9450	−0.0483
2	0.9995	−0.0003	2	0.9412	−0.0521
3	1.0002	0.0004	3	1.0720	0.0787
4	1.0003	0.0005	4	1.0713	0.0780
5	0.9992	−0.0006	5	0.9410	−0.0523
6	1.0000	0.0002	6	1.0520	0.0587
7	0.9998	0.0000	7	0.9498	−0.0435
8	0.9996	−0.0002	8	1.0640	0.0707
9	1.0003	0.0005	9	0.9426	−0.0507
10	0.9994	−0.0004	10	0.9407	−0.0526
11	0.9993	−0.0005	11	0.9503	−0.0430
12	0.9995	−0.0003	12	1.0501	0.0568
Mean	0.9998			0.9933	
Standard deviation		0.0005			0.0609
$0.0004 \ll 0.0609$					

* the difference between the results and the mean value

Factors influencing the uncertainty during pipetting of real object

- Viscosity of the liquid;
- Presence of substances that change the density of the liquid;
- Contamination of the inner surface of the glass;
- Change of temperature of the liquid in relation to the temperature in which the calibration was conducted.

Table 8.4 shows the results of weighing the distilled water and the solution of NaCl, 2% (m/v). A pipette with the nominal volume of 25 mL was used. The mass of the liquid was determined as the difference between the mass of the vessel and pipetted portion of the liquid. In both experiments, 12 consecutive measurements were conducted, and the results are shown below.

BE AWARE! When pipetting liquids and solutions other than the distilled water (e.g., the solution of NaCl), the uncertainty (expressed as the standard deviation) can come to a value much higher than the uncertainty obtained during calibration (as provided in the certificate of the pipette).

Example 3 In spectrophotometric measurements, a known and well-defined Lambert-Beer law is used. The uncertainty of the measurement of absorbance depends on the following parameters:

- Repeatability of consecutive measurements;

Table 8.4 Results of weighing various liquids (distilled water and solution of NaCl)

No	Distilled water		No	Solution of NaCl	
	Mass, g	Δ*		Mass, g	Δ*
1	25.011	0.067	1	25.602	0.810
2	25.092	0.148	2	25.601	0.809
3	25.103	0.159	3	25.603	0.811
4	24.895	−0.049	4	25.704	0.912
5	25.004	0.060	5	24.012	−0.780
6	24.000	0.056	6	25.520	0.728
7	24.898	−0.046	7	24.498	−0.294
8	24.896	−0.048	8	24.426	−0.366
9	24.894	−0.050	9	24.407	−0.385
10	24.893	−0.051	10	24.103	−0.689
11	24.792	−0.152	11	24.010	−0.782
12	24.855	−0.089	12	24.015	−0.777
Mean	24.944			24.792	
Standard deviation		0.01			0.74
$0.01 \ll 0.74$					

*the difference between the results and mean value

- Drift of the basic line of the photometer;
- Deviation from the Lambert-Beer law;
- Assumed method of rounding the result.

In the case of measurements conducted for 'ideal' samples, for example, distilled water, commonly the uncertainty of measurement of absorbance is very small. In reality, for the solutions of compounds of interest, the measurement of absorbance can be influenced by the matrix component, which can significantly increase the measurement uncertainty.

Factors that influence the uncertainty of measurements of absorbance
- Absorbance of light by other components of the solution;
- Influence of the matrix on the stability of the colorful complex;
- Instability of the colorful complex;
- Presence of suspension in the solution.

BE AWARE! When measuring the absorbance for the solutions, the uncertainty (expressed as the standard deviation) can come to a value much higher than the uncertainty obtained during calibration (as provided in the certificate of spectrophotometer).

Determination of nitrates (standard procedure ISO 6777:1984)
Description of the measuring procedure: after adding the reagents, in 20 min, a colorful complex is created, with a durability of 2 h, which means that the measurement can be conducted about 1.5 h after a stable color is obtained.
In laboratory practice: depending on the components of the matrix, a gradual decrease of the absorbance can be observed, of about 15% of the value noted directly after the color stabilizes. Moreover, in the first 15 min (when the measurements are most often conducted), the absorbance decreased by almost 6%.

Example 4 In chemical measurements, an essential part of the analytical procedure is the sample preparation. In most cases, it is not possible to describe the influence of each of the step of the sample preparation process in the form of a mathematical equation (modeling). It is also not easy to assign numerical values to the uncertainty introduced by those steps.

In practice, to assess the influence of sample preparation, the recovery test is used. This can be executed by use of the reference materials, similar in their properties to the tested samples, or by adding a known amount of analyte to the investigated sample. Thus, the uncertainty of recovery should be defined, which is not always easy, and does not always reflect the real behavior of the tested object.

Factors influencing uncertainty of recovery
When using reference materials
- Disparity between the form (physical state, granulation, uniformity) of the test sample and the reference material;
- The chemical and physical form of the analyte in the reference material;

When using test samples with the addition of the standard
- Disparity between the chemical form in which the addition is introduced to the test sample, and in which it is present in the sample;
- The step of the analytical procedure in which the standard is added;
- The time used for establishing the chemical equilibrium between the analyte added and originally being present in test sample.

BE AWARE! The uncertainty of the recovery for test samples with the addition of the standard can be much higher in comparison with the recovery of the reference material.

8.20 Advantages and Limitations of the Modeling Procedure for Evaluation of Uncertainty

To summarize, it is worth emphasizing both the advantages and the limitations of using modeling in the evaluation of the uncertainty of chemical measurements.

Advantages:

- Uniform and commonly accepted method of evaluation of uncertainty;
- Analytical procedure can be described in the form of a mathematical equation;
- Enables the optimization of the analytical procedure;
- Accepted by international organizations.

Limitations:

- Analytical procedure is often not easily modeled;
- Not all uncertainty contributions are easily quantified;
- Procedure requires a good knowledge of mathematical statistics;
- Procedure requires a detailed analysis of the measuring procedure;
- Often leads to the underestimation of the uncertainty.

8.21 Experimental Methods of Evaluation of Uncertainty

The modeling method requires the identification and clear determination of all constituents of the measuring procedur—those that influence the measuring result and its uncertainty. It means that all stages of the procedure are well defined, their influence on the result is known and it is possible to estimate the value of uncertainty for all constituents.

In chemical measurements, where the measuring procedure comprises many stages, it is not always possible to display a full mathematical model. The measurement result often depends on parameters that are hard to predict. The laboratory practice shows that some values determined during the measurements reflect the influence of a few sources of uncertainty at the same time.

Evaluation of uncertainty: useful data

- Data collected during the validation of the measuring procedure in the laboratory;
- Data collected during quality control (i.e. control charts) in the laboratory;
- Data from inter-laboratory comparisons (inter-laboratory validation);
- Data from proficiency testing.

Validation of measuring procedure in the laboratory

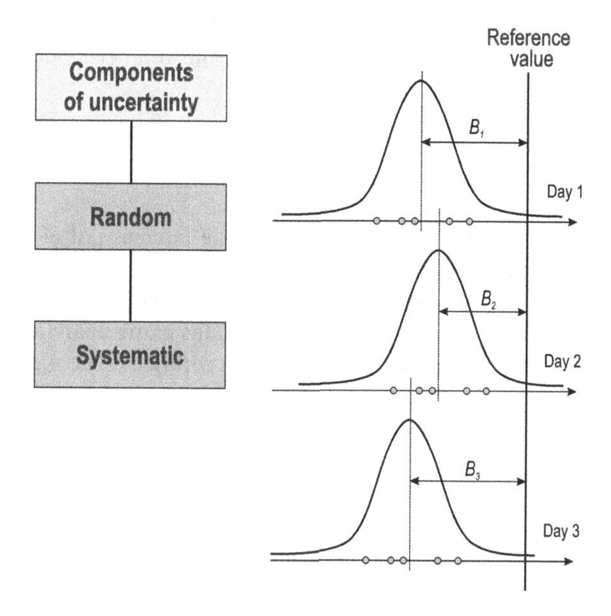

Fig. 8.3 Uncertainty components of random and systematic origin related to the repetitive measurements

Whether the selected analytical procedure is fit for purpose is assessed through its validation performed in the laboratory. It is therefore essential to carefully characterize the performance parameters of the analytical procedure, in order to compare whether they are relevant for a specific, intended use of the results. Within the validation, measurement data are collected and can be used later on to support the evaluation of uncertainty. In order to use those data, careful planning of the experiments is needed. It is important that collected data reflect the predicted (possible to assess) variability of the type of samples (e.g., whole, semi-skimmed and skimmed milk) and the variability of the content of the analyte in the tested objects as well as the variability of the measuring conditions occurring in a given laboratory.

The main components of the uncertainty are those estimating the influence of all systematic effects (u_{sys}) and random effects (u_{rand}).

Figure 8.3 shows the dispersion of results on a given day, and in consecutive days. In this case, the combined standard uncertainty u_c can be written as:

$$u_c = \sqrt{(u_{sys})^2 + (u_{rand})^2} \quad (8.21)$$

The uncertainty that estimates the influence of all systematic effects (u_{sys}) can be delivered from the data:

- Of the test sample;
- Of the test sample enriched with the known quantity of the analyte;
- Of the reference material with the certified content of the analyte;

– From the assessment of the results of inter-laboratory comparison (with the use of CRM).

The uncertainty that estimates the influence of all random effects (u_{rand}) can be delivered from the data:
– Of test samples (stable over the time of collecting data);
– Of a blank sample referring to the entire analytical procedure;
– Of a reference material with a certified content of the analyte.

The procedure used for the evaluation of uncertainty from validation data:

1. Defining the measurand (analyte, sample type);
2. Calculating recovery value R and the assigned uncertainty;
3. Calculating laboratory reproducibility;
4. Evaluating the standard uncertainties of the systematic and random effects;
5. Calculating the combined standard uncertainty: using the law of propagation;
6. Calculating the expanded uncertainty with the selected coverage factor $\boldsymbol{k}$.

Data from the control charts

When the evaluation of uncertainty is performed within validations, useful data are collected over the relatively short period of time in which the validation was conducted. In laboratory practice, validation experiments are often conducted by one employee, with the use of one set of glassware, reagents and instruments. This means that the dispersion of data reflects the uncertainty related to the random effects occuring over the validation time.

In the long-term, a larger dispersity of results can be expected, due to the higher possible variability of the measurement conditions. With this in mind, whenever possible, it is recommended that data (e.g., standard deviation) from control charts is used. When the control charts were completed over a long time or a large enough amount of data was collected, it can be assumed that standard deviation reflects the intra-laboratory reproducibility, encompassing the changeability of the following parameters:

– Personnel conducting the test (when more than one person is authorized to conduct the test);
– Environmental conditions (temperature, humidity);
– Equipment (stability, repeatability);
– Reagents (stability, purity);
– Standards (uncertainty of certified value, matrix).

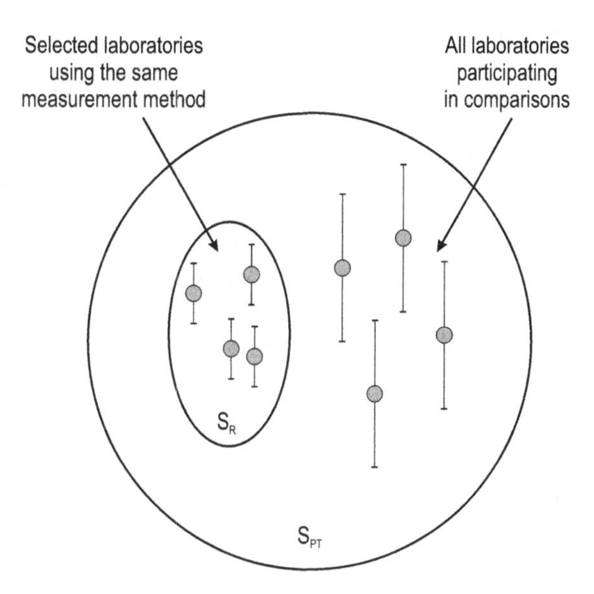

Fig. 8.4 Example of the discrepancy of the results between various laboratories

It is obvious that a given measuring procedure should be used and that the measurements should to be conducted under the conditions used for the routine test samples.

Inter-laboratory comparisons

Data from inter-laboratory comparison (ILCs) reflect the spread of results of measurements conducted in different laboratories. The variability of the conditions between laboratories is usually bigger than in a single laboratory. The inter-laboratory reproducibility can, therefore, be a good source of information used for estimation of the uncertainty, under the condition that the tested objects are similar to the routine samples tested in a laboratory. In cases when different techniques or even different analytical procedures were used, it can be advisable to calculate the standard deviation for the selected set of results submitted by those laboratories that used the same technique/analytical procedure.

A number of the providers of ILCs during the processing of the results include statistical data for selected laboratories in the report that use the same measuring method (Fig. 8.4). However, this is only possible under the condition that the number of such laboratories is high enough to make a statistically sound evaluation.

The value of the standard deviation of ILC (s_{ILC}) is assumed to be the standard uncertainty ($u_{c(ILC)}$)

$$s_{ILC} = u_{c(ILC)}$$

Thus, expanded uncertainty is calculated as $U = 2\ u_{c(ILC)}$

Proficiency testing

The main objective of the proficiency testing (PT), executed via ILC, is to evaluate the competences of a laboratory; thus the results of PT tests can also be a good source of data for the evaluation of uncertainty of measurements conducted in a given laboratory. The use of the results of the PT is subject to the same conditions as the use of any other results of ILC. The only limitation, described in various documents (e.g., EA-4/16 'EA Guidelines on the Expression of Uncertainty in Quantitative Testing') is that in some cases the proficiency testing is not conducted regularly enough so that it could be possible to collect a representative set of data.

Another limitation with the use of the data from PT for evaluation of the uncertainty is that the tested object does not sufficiently reflect the properties of samples that are routinely tested in a given laboratory. In that case, the uncertainty budget should include the predicted differences in the behavior of the object tested in the laboratory with regard to the routine samples. This should be also considered when the uncertainty varies over the concentration range for which the procedure is applied in the given laboratory.

8.22 Conclusions

In practice, four approaches for evaluation of uncertainty could be applied. The particular approach can be selected, depending mainly on the purpose of conducting measurements and depending on the availability of source data. For this reason, the use of a combined approach is considered to be the most effective.

It should be highlighted, that apart the different sources of uncertainty components, the general process in always the same for all four approaches. The process always starts with the specification of measurand, which means there is a necessity to clarify and define the quantity intended to be measured.

Measurand: the quantity intended to be measured.
Clause 2.3; ISO/IEC Guide 99

Whenever the measurand was defined, the measurement procedure should be selected and described in the form of the mathematical function (model equation). The measurement procedure should be selected considering the measurand of interest as well as target uncertainty and available resources. The measurement function should first of all ensure the proper calculation of the final results and should be updated when all possible sources of uncertainty are identified. In this respect, understanding the entire analytical procedure and especially all possible effects that could affect measurements is a crucial part of the evaluation of uncertainty.

The next step is related to the quantification of the uncertainty components and is different for all possible approaches, as described above.

Depending on the data available as well as purpose of the measurement, different approaches for the evaluation of measurement uncertainty can be used.

In the modeling approach, the components of uncertainty are quantified individually, and the uncertainty refers to a particular measurement result.

The single laboratory validation and data from control charts uses data from the given, single laboratory, uncertainty components are grouped into several major influences.

In the ILC/PT approach, uncertainty source data arrive from several laboratories, using the selected measurement procedure, thus reflecting uncertainty under reproducibility conditions.

Uncertainty by different approaches		
Modeling	Uncertainty components are quantified individually.	Refers to an individual result
Single laboratory validation and QC	Uncertainty component are grouped	Refers to results obtained in a single laboratory (repeatability conditions)
ILC/PT	Uncertainty estimate for any laboratory performing a given measurement	Refers to results obtained in several laboratories (reproducibility conditions)

Comparison of different approaches for the evaluation of uncertainty

Modeling

- Requires extensive knowledge and competences;
- Labor-consuming;
- Requires a good knowledge of the statistics;
- Often leads to underestimation of the uncertainty;
- Allows in-depth understanding of the given measuring procedure and highlightts parameters that have the biggest influence on the uncertainty.

Single-laboratory validation

- Justified for the routine laboratories;
- Requires gathering of a lot of data;
- Does not require additional effort;
- Allows for a complex assessment of uncertainty for a measuring procedure in the conditions of changeability in a given laboratory.

Inter-laboratory validation (e.g., ILC/PT)

- Does not require additional measurements to be conducted or a detailed knowledge of the measuring procedure;

- Is an estimation for the changeability of conditions in different laboratories;
- Can be an indicator of the typical values of uncertainty (as an approximation).

Expressing the uncertainty: general remarks

- For low concentrations (close to the limit of determination), use absolute values—at this level, the uncertainty does not depend on the concentration;
- For high concentrations, use relative values—for high concentrations, the uncertainty is approximately proportional to the concentration.

The meaning of the uncertainty

The uncertainty budget comprises many components that influence, to a various degree, the combined uncertainty. In laboratory practice, which of the components have a significant influence on the combined uncertainty and which are neglected can be assessed.

When constructing the uncertainty budget, it is recommended to carry out a preliminary assessment of all identified components, so as to select those whose contribution is most significant. It can be concluded that those components whose value is not higher then one fifth of the biggest component will only have a 2% share in the combined standard uncertainty.

BE AWARE! Always consider the number of components of the same kind. If they are omitted individually, this can lead to underestimation of the combined uncertainty.

Uncertainty and decision-making

The knowledge on uncertainty values is essential in all situations where the result is close to the decision limits; for example, the highest acceptable concentration of a substance contaminating the sample or the lowest acceptable level of a given component (e.g., biologically active substance in a pharmaceutical formulation). In such cases, the value of the uncertainty assigned to the result can have a significant meaning in decision-making, hence attempts are made to obtain the smallest possible uncertainty. In practice, situations arise like the ones shown in Fig. 8.5.

Cases 1 and 4

The result of the measurement with the assigned uncertainty does not include the decision value in its range. Assuming that the presented case concerns the highest acceptable concentration of a given substance in the tested object, two scenarios can be distinguished in which the decision-making body can make a clear-cut decision:

1. The result with the assigned uncertainty is above the permissible content.
The sample does not meet the requirements.

4. The result with the assigned uncertainty is below the permissible content.
The sample meets the requirements.

Fig. 8.5 Result with accompanying uncertainty close to the decision limit

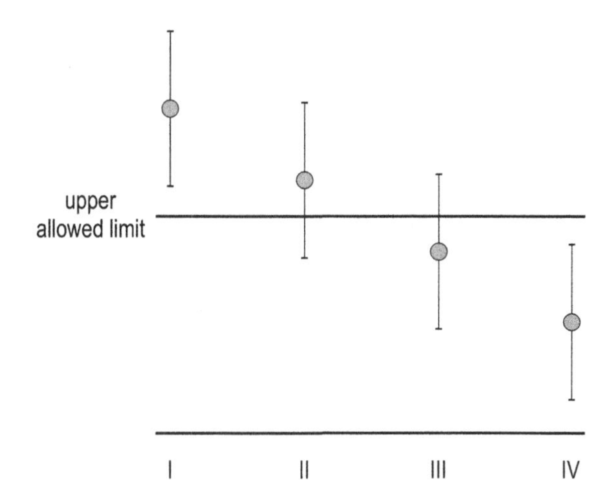

Cases 2 and 3

The measurement result with the assigned uncertainty includes the decision value in its range. In both cases, it is therefore important to define the decision-making criteria beforehand.

If the criterion will be the mean value without the assigned uncertainty, then case 2 should follow case 1 (the sample does not meet the requirements), and case 3 should follows case 4 (the sample meets the requirements).

If the criterion will be the result with the assigned uncertainty, then for case 2 and 3, that decision will mean that the sample meets the requirements. In both cases, however, it will be necessary to establish beforehand what legal consensus applies to it, how often can such a situation occurs and what further actions will be taken by the decision-making body. More information on the subject of the meaning of the uncertainty value in decision-making can be found in Ellison S.L.R. and Williams A. (Eds). Eurachem/CITAC guide: *Use of Uncertainty Information in Compliance Assessment*, First Edition (2007); available from www.eurachem.org.

Chapter 9
Managing the Quality Systems

Quality is a set of features and characteristics of a product or a service, conditioned by the fact that they meet set and expected requirements.

The term 'quality' has accompanied humankind since the beginning of evolution. Initially, it was a philosophical term, connected to the understanding of the matter surrounding humans. The term 'quality' first emerged in Aristotle's writing as a metaphysical category counterposed to 'quantity.' In his consideration on the essence of beings, Aristotle distinguished two components: the form (that is the general properties of things) and matter (the individual properties of a given thing). In the Aristotelean philosophy, 'quality' is the term that describes the definiteness of the corporeal substance and is evidenced by a specific assignment of a matter to form. In the philosophy of Rene Descartes, a dualistic understanding of qualities can be found: primary qualities, those that are already in the object; and secondary qualities, those emitted by the object.

Nowadays, the understanding of 'quality' goes far beyond just a philosophical term related to the perception of matter and has become a socioeconomic term related, to a great extent, with human activity. Taking into consideration the understanding of quality as a certain kind of property of the product, the term can be examined on a few dimensions: economical, technical and social. Recognizing that the measurement result is 'a product' delivered by the laboratory, the quality of the measurement matters on all of the aforementioned dimensions. It is needless to say that correct and sound decisions can be taken only on the basis of reliable results.

Quality is not objective, it depends on the criteria used to evaluate a given object or phenomenon. Quality also depends on our needs—that is, the person who uses a ware or service must define what they expect.

The important step forward conducive to the evolution of the term 'quality' in the direction related to the effects of human activity was the industrial revolution that

E. Bulska, *Metrology in Chemistry*, Lecture Notes in Chemistry 101,
https://doi.org/10.1007/978-3-319-99206-8_9

began in England at the turn of the eighteenth and nineteenth century. It led to the shaping of the structures of industrial production and the use of steam engines caused a sharp increase in the quantity of produced wares. The increase in the production scale also gave a start to actions directed at controlling the production process in order to limit the losses of the producers—those connected to the unsatisfying quality of the products (so-called Taylorism). This covers technical control, development of organizational schemes, methods and scope of quality management, conceptions of managing the quality evolved in the direction of 'total quality,' encompassing not only the production line but the functioning of the entire company. Nowadays, it is believed that the management system encompassing the functioning of the organization, as well as the technical aspects, is the best scheme of quality management.

A high quality results from chemical measurements requires the laboratory to be well organized; proper infrastructure and technical competencies of the personnel are also necessary.

The development of the quality management strategy was tightly connected with the increasingly common use of statistical methods, and a significant step was the appearance, in 1931, of the book by Walter Shewhart, *Economic Control of Quality of Manufactured Product*, in which the term statistical control of processes appeared. Shewhart assumed that every process is laden with a certain variation and because of that it should be constantly controlled statistically in order for it to be possible to trace trends and prevent deviation. For the purpose of analyzing the changeability of the process, Shewhart proposed control charts for managing quality, nowadays called also as Shewahart's charts. In the course of conducting of the control charts, it is assumed that each process should stay within set tolerance limits.

Laboratories use control charts to monitor the direction of changes of selected analytical parameters of the measuring procedure used.

9.1 Quality Management System

Quality Management System (QMS) encompasses Quality Management (QM), Quality Assurance (QA) and Quality Control (QC), as described in the ISO 9000 standard 'Quality Management System—Fundamentals and Vocabulary.' The organization management system as such, including relevant requirements, is described in the ISO 9001 standard 'Quality Management System—Requirements,' a document designed to be used in any institution (not necessarily a laboratory), meaning that there is confirmation that it is performing in accordance with the requirements of that document.

Standards from ISO 9000 series state the management rules and the general rules of quality management; however, they do not refer in detail to specific technical areas. That lack of trade reference has resulted in the need to develop requirements tailored to certain technical sectors. Because the ISO 9001 standard does not contain in its scope the technical aspects related to the performance of the test and calibration laboratories, a document was drafted, designed especially for the needs of evaluating the technical activity of test and calibration laboratories: the ISO/IEC 17025 standard *General Requirements for the Competence of Testing and Calibration Laboratories.* It is worth noting that the first edition comes from 1999—that is, before the ISO 9001:2000 standard was issued. The ISO/IEC 17025 standard replaced two previously used documents, ISO/IEC Guide 25 and EN 450010, and at the same time, it referred to the ISO 9001 and ISO 9002 standards issued in 1994. The next, updated edition of the ISO 9001:2001 standard (replacing the two mentioned above) has led to the need to update the ISO/IEC 17025 accordingly. The second issue of the ISO/IEC 17025:2005 was valid until the end of 2017, when the ISO/IEC 17025:2017 was approved.

The withdrawn ISO/IEC 17025:2005 standard comprises five chapters, out of which the most important are Chaps. 4 and 5. Chapter 4 includes the requirements regarding the management of the laboratory, and Chap. 5 includes the requirements regarding the technical performance, including proper equipment for laboratories, the development of measuring procedures, ensuring proper environmental conditions (e.g., temperature, humidity), and so on.

The current edition of ISO/IEC 17024:2017 supersedes the previous version. A new structure has been adopted to align the standard with the other existing ISO/IEC conformity assessment standards such as the ISO/IEC 17000 series on conformity assessment. It also covers technical changes, vocabulary and developments in IT. It should be highlighted that the major requirements for the competence of testing and calibration laboratories has stayed as it was; however, the scope was extended to also cover sampling associated with subsequent calibration and testing. The new expectation is a focus on the process approach, which also matches ISO 9001 (quality management), ISO 15189 (quality of medical laboratories) and ISO/IEC 17021-1 (requirements for audit and certification bodies). Moreover, the requirements related to risks and opportunities are added. Finally, the new edition focuses on information technologies and incorporates the use of computer systems, electronic records and the production of electronic results and reports.

All revised standards put the emphasis on the processes instead of the detailed description of its tasks and steps. As already mentioned, the updated standard focuses on information technologies and incorporates the use of computer systems, electronic records and the production of electronic results and reports. The concept of risk-based evaluation of the laboratory activities is newly introduced. Last but not least, the terminology has been updated to be more in accordance with other standards and technical documents.

9.2 Quality Assurance and Quality Control

In general, the quality management system combines a management of the organization as such as well as quality assurance (QA) and quality control (QC). QA is a set of organizational and technical actions, which take into consideration the scope of the laboratory's activity and, in accordance with the knowledge available, allow creation suitable working conditions for a given laboratory. Works in the scope of QA are very broad and encompass all requirements of ISO/IEC 17025. QA in the given organization encompasses a number of tasks; for example, flow of documents and records; customers service; training of staff; laboratory infrastructure; validation of the analytical procedures, as well as metrological services (e.g., calibration).

Quality assurance (QA) and quality control (QC)
Quality assurance
A set of planned organizational and technical actions aimed at obtaining reliable measurement results
Quality control (sometimes described as 'quality management')
Systematic control of selected parameters of the measuring procedure and confirmation of their compliance with previously specified criteria

The laboratory also has to analyze its actions, usually through internal evaluation (audits) and/or on the basis of client feedback, and in the case of situations that are not in compliance with the requirements, it needs to undertake proper actions. In a well-organized laboratory, the risk of delivering incorrect results to the client should be relatively small, thanks to systematically used control mechanisms (i.e., QC). QA also encompasses a number of technical actions, tailored to conducted measurements, including:

- Use of validated measuring procedures;
- Proper supervision of equipment, including calibration of those measuring devices that influence the result of the measurement;
- Proper sample storage and proper procedure of the preparation of studied objects, including the process of acquiring a few parallel portions of the same sample;
- Taking care of proper environmental conditions in the laboratory (e.g., temperature, humidity).

Apart from that, it is extremely important to specify the requirements regarding the quality indicators and the frequency of their control. That area of laboratory activity is counted among the actions connected to quality management—that is, a process of systematic control in order to obtain information (with the help of the previously selected indicators) regarding whether the systematic actions are still effective.

The laboratory should have procedures for control management in order to monitor the reliability of the tests and calibrations conducted. Obtained data should be

Table 9.1 List of items related to the quality management

Quality management	
Quality assurance (QA)	Quality control (QC)
– Quality management system in place – Proper environmental conditions – Competent personnel – Training program – Calibration procedure – Planning of calibration – Validated procedures – Assuring traceability – Knowledge of uncertainty – Presenting the results – Control of records – Quality of standards and reagents – Implemented system of internal QC – Implemented system of external QC	– Testing of reference materials – Testing of control samples – Testing of blank samples – Testing of spiked samples – Testing of archive samples – Simultaneous testing of the same object – Participation in ILC (Inter-Laboratory Comparison) – Participation in PT (Proficiency Testing)

recorded in a way that enables the tracking of the trends and, if it is possible, statistical techniques should be used in order to review the results.

The monitoring should be planned and subjected to inspections, and can encompass, but is not limited to the following:

- Systematic use of certified reference material (CRM) and/or internal QC with the use of laboratory control samples;
- Participation in inter-laboratory comparison (ILC) or in proficiency testing (PT);
- Repeated measurements for tests samples;
- Correlation of results referring to different properties of the objects.

The data collected within the process of QA/QC should be analyzed so as to determine any deviation from the previously set criteria; if this is present, actions should be taken to correct the problem.

The actions connected to QA stem from the need to assess the risk of making a mistake during testing and should be prepared so that the risk of giving the client the incorrect result to the client is the smallest possible.

The actions connected to QC should enable the behavior relating to measuring systems to be tracked in the laboratory and can be the 'early warning' tool in cases when situations arise that can lead to errors (minimizing the risk).

It is important for the procedures of quality management used in practice to be suitable for the kind and size of the work, which means that in each case, the laboratory should tailor a set of proper tools to its needs that will ensure the quality of the results (Table 9.1).

To summarize, quality management in a chemical laboratory performing tests or calibrations, comprises two complementary areas:

- Quality assurance (QA) is a set of procedures that creates a system ensuring the creation of the best possible conditions tailored to the type of tests conducted;
- Quality control (QC) is a systematic supervision over technical activities, encompassing the control of selected parameters (quality indicators).

Systematic evaluation of those parameters will make it possible to ensure that the system works effectively, and possible deviations will be noticed in time so that they can be avoided and proper corrective actions could be taken.

9.3 Quality Control

In a laboratory, in which the requirements regarding the quality of results have been specified, it is necessary to employ a systematic evaluation of performance. The commonly accepted approach used for effective QC, includes a systematic evaluation of selected quality indicators via comparison with the previously defined requirements. A well-designed system of QC ensures that the risk of reporting incorrect results is minimized via an early warning methodology.

QC can be of internal as well external means. External quality management is, more than anything, the participation of the laboratory in inter-laboratory comparison and/or in proficiency testing. In that case, the laboratory receives a sample of an unknown value of a given property (e.g., the substance's concentration and/or its identification), conducts a test, sends the results to the ILC/PT providers and finally is evaluated on the basis of the degree of compliance with the reference value.

As long as the laboratory regularly participates in ILC/PT, it is also possible to use control charts to monitor the trends. Monitoring the quality of results is made on the basis of the value of the performance scores; for example, z-, z'-, zeta and En scores. More details on the external QC, manly via participation in ILC/PT will be given in Chap. 10.

9.4 Quality Measures

As previously mentioned, the scope of QC, meaning the selection of quality measures as well as the frequency in which they are checked, should be designed so as to fit to the kind and number of the tests performed. The proper selection of the study items involved in the measuring procedure is extreamly important and should undergoes the entire procedure used in a given laboratory. Thus, the measurement results should be recorded so as to evaluate whether the analytical system in a single laboratory is working correctly. It is known that in practice, slight variation of the response occurs; therefore, the purpose of the quality control is to monitor whether the fluctuation of results stays within the acceptable range. For this purpose, it is worthwhile to use the statistical methodology, in order to use distributions and probability functions.

Internal QC encompasses the evaluation of results obtained in the laboratory, involving the entire analytical process. The commonly accepted approach is the use of control charts, where the reported values are plotted on the chart so as to evaluate whether the measurements are performed within the given limits. All control samples (e.g., blanks, control samples, standards), should run as routine test samples in exactly the same conditions, using the same analytical practice. It is convenient to record the results of measurements directly on the control charts, since it allows for the visual evaluation of any trends and/or immediate confirmation if the value of the given property is with previously established limits.

Control charts are therefore a valuable source of information on the quality of the measurement; they allow the stability and trends to be tracked.

9.5 Routine and Blind Samples

In the case of internal and external quality control, samples may enter the analytical process as known control samples, which means that the analyst is aware it is a control; or as a blind samples mixed with other routine test samples, which means that the analyst cannot identify the control sample before performing measurements. Both approaches have their advantages and limitations; however, the main aim is to process the control samples under the everyday conditions.

9.6 Internal Quality Control

A laboratory should design a program of internal quality management for every type of test (method, sample type) and such a program should be carefully planned and documented. The selection of control samples is also important; they should allow information on the behaviour relating to the measuring system to be obtained. Control samples should be stable and available in such a quantity that they could be used for QC purposes for a long time. Various samples can be used as a control items; they should be selected so as to be fit for purpose. The most commonly used in laboratory practice are matrix CRM, reference material (RM), blank samples, routine test samples or spiked routine test samples.

> When the result obtained for the control sample is within the previously set limits, it is assumed that analytical system is stable and under control.

9.7 The Frequency of Internal Control

There are no explicit and general rules regarding the frequency of the use of control samples in a given measurement series. This should always be evaluated individually, regarding the type of routine test samples, the number of items analyzed, and previous experience regarding the stability of the analytical system. The analyst is the one who should decide on the proper sequence of measuring the control samples. It is important that the measurements of control samples are executed so as to accompany the routine test samples.

When tailoring the frequency of the internal control, it is worth taking the following under consideration:

- At least one control sample should be used in a given measurement batch;
- The order of control samples in the series of test samples should be random, bearing in mind the recommendation that the control should be executed at the beginning and at the end of the measurement batch;
- In cases where the content of the determined substance varies significantly for the routine test samples, it is recommended that two or more control samples be used for different levels of concentration;
- In routine measurements, often at least one control sample is used in the batch of 20 test samples (certainty level 5%);
- In screening tests, often one sample is used for every 40 or 50 laboratory samples.

9.8 Control Samples

As mentioned previously, items used as control samples could be of different origin, including matrix RMs, chemical standards, laboratory RMs, routine test samples, blanks, to list some of the most commonly used. In a given measurement procedure, either the single control sample can be selected, or the set of control samples are used to monitor various features of the measurement process.

Bellow, the most commonly used control samples is described.

9.8.1 Certified Reference Materials and Chemical Standards

Those are the most commonly used for checking any systematic effects as well as for estimating the random variation of the measuring system. Chemical standards of known purity could be prepared by laboratory or could be obtained from the suppliers of chemicals; CRMs are usually supply by the qualified producers, both should be of known purity and known content of the substance of interest.

- Pure substance (material with a certified content of the substance)
- Certified reference material (CRM)

Both types of control samples are very useful for the evaluation of random variation and to obtain estimation of any systematic effect, as well as for monitoring trends over time. The control samples prepared from pure chemical substances are expected to be within the concentration range close to that of routine test samples, but are not always able to mimic the composition of the matrix. Although CRMs, purchased from the qualified producer, seem to be the best chose as a control samples, they are not always available for all possible analytical situations. It is important to stress that CRMs are always of better homogeneity; thus the standard deviation of measurements are always smaller compared with that obtained for routine test samples. Thus, the natural samples of sufficient stability could be a good choice for the evaluation of the typical spread of results for repetitive measurements.

9.8.2 Routine Laboratory Samples

Routinely analyzed laboratory test samples, whenever they exhibit sufficient stability over time, could be very useful for the estimation of the random variation of the results. They are of exactly the same matrix and exactly the same homogeneity; thus, to the best extent, they mimic the routine analytical situation. When they are used, the best estimation of repeatability and reproducibility can be obtained; thus, they could be used for the day-to-day monitoring of the correctness of the analytical procedure. However, the use of routine laboratory test samples, as obtained, do not provide sufficient information about the accuracy of the measurement results. For this purpose, the spike samples can be used instead.

When the known amount of standard is added to the sample, the increase in the analytical signal can be used for the evaluation of the matrix effects. This is done mostly by monitoring the recovery for the given type of matrix. The most valid information of the influence of matrix is possible when the spike sample is subjected to the same analytical processing as the sample. The analyst should be aware, however, that the spiked standard cannot always undergo the same binding as the originally occurring substance in the natural sample.

9.8.3 Blanks

Blanks are those samples that do not contain the substance of interest. Blank samples could mimic the standard solution of pure substance (e.g., for standard solutions of cadmium in 0.5 mol/L HNO_3, a blank will be a solution of 0.5 mol/L HNO_3), or they could be a test sample where the content of substance of interest is below the detection limit.

Blanks can be used for the evaluation of the purity of reagents used in sample preparation (reagent blank), for the determination of the limit of detection and limit of determination.

9.9 Principle of Quality Control

Quality control is used to monitor the stability of the analytical system; thus, various tools can be used accordingly. Several measurements can be used to give the relevant information:

- Measurement of blank samples;
- Measurement of chemical standard or CRM;
- Measurement of test sample—as it is or spiked with standard;

When selecting the type of control sample, various approaches can be used, including:

- Measurement of repeat samples;
- Measurement of blind samples;
- Measurement of test samples.

NOTE! The number of replicate measurements for control samples should be the same as used for the routine test samples. It is important for the samples used in the quality management (i.e., standards, CRMs, RMs, laboratory samples, blanks) to be analyzed in the same way and in the same conditions as the laboratory samples.

9.10 Control Charts

Control charts are the graphical presentation of data obtained for the measurements performed for the control samples, representing the flow of the analytical process over time. It is essential that current measurement results are recorded on the chart in order to visually evaluate it with regards to previously established limits or, when they are considered as provisionary, to allows those limits to be established. Control charts are graphs that plot data in time-ordered sequence. Most control charts include a center line, an upper control limit, and a lower control limit. The center line represents the mean value, and the control limits represent the process variation. By default, the control limits are drawn at distances of 3σ above and below the center line.

When processing the large set of data, their statistical distribution should be considered. In most cases, results of subsequent measurements display a certain distribution around the mean value, and if they are subject to regular distribution, then they will arrange themselves symmetrically in relation to the mean value (the measure of

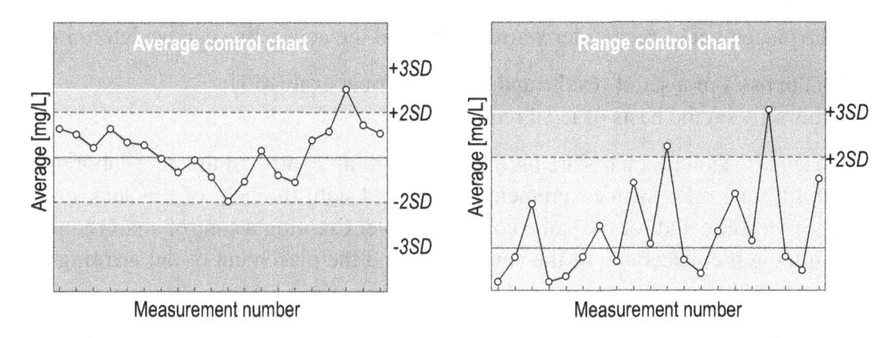

Fig. 9.1 Two-sided and one-side control charts

the distribution will be the value of standard deviation). Taking into consideration a normal distribution, it can be assumed that about 95% of values will fall within the range $\pm 2\sigma$, and 99.7% will fall within $\pm 3\sigma$ of the mean value. It other words, it is unlikely (5% probability) that the obtained result will be outside the 2σ and very unlikely (0.3% probability) that we will obtain a result outside the 3σ range. If results obtained over a longer period of time meet the set criteria, then we can assume that other results should behave the same way, otherwise we can treat that as a warning that there was a change in the measuring system. The main objective of control charts is the evaluation of the measurement system stability and to remark on possible deviations. Graphic charts allow for visual evaluation of the set of numerical values—that is, the measurement results obtained for a given control sample.

Control charts allow for a graphic illustration of two types of variations:

- Random variation;
- Systematic trends.

9.10.1 Types of Control Charts

In test and calibration laboratories, the following control charts are used:

- Shewhart's charts (for measurement results, mean value, range, recovery);
- CuSum chart.

Shewhart's control charts have a central line and lines placed symmetrically on both sides of the central line; warning lines placed at the distance of 2 σ from the central line and lines indicating the need to take actions at the distance of 3 σ from the central line. In special cases, for example, to control the range between two measurements, one-sided charts are used as well (Fig. 9.1).

The placement of the central line can be determined:

- As the mean value for a measurement series for the control sample;
- As a reference value (from a certificate or a recognized reference value).

The placement of the border warning line and the action line can be determined:

- On the basis of a set of results and their statistical evaluation;
- Arbitrarily on the basis of legal requirements.

Shewhart's control charts are used for the ongoing control of the inter-laboratory variability of results of measurements conducted with the help of the same measuring procedure and for the same control sample. Creating a control chart requires determining the placement of the central line and the placement of the warning (2) and action (3) lines.

9.10.2 Shewhart's Chart

Figure 9.1 shows an example control chart of the Shewhart type. The control chart usually holds 20 or, less often, 40 measurement results. The nominal value is placed on the chart before the start of the measuring series.

In the case when the value for the central line is not an arbitrarily accepted reference value (nominal, normative), the mean value and the standard deviation are calculated after the first control chart for a given control sample has been filled.

9.10.3 Recovery Control

Control charts can also be used to evaluate the recovery when a known amount of the RM is added to the real samples. In an ideal scenario, the recovery rate should be 100%, whereas the variability of results obtained for subsequent measurements (standard deviation) should be a component of the uncertainty budget. In reality, it often happens that the recovery is smaller or bigger (depending on the type of interference) than 100%; in such situations, the information obtained on the basis of the gathered results allows the recovery value (the systematic component) to be evaluated in addition to evaluation of the results spanning the mean value of recovery (the random component). Both of those values must be included when giving the result.

The variability of results is always a component of the uncertainty budget, whereas the method load can be a component of the uncertainty budget or it can be included in the calculation in the form of a recovery factor.

Example

Case 1: both components (systematic and random) are included in the uncertainty budget.

The content of methylmercury in tuna tissue has been determined. On the basis of the test of enriched samples (after their incubation), the recovery value was determined at the level of 97% + 4%. The value of the measurement uncertainty, stemming from the requirements of the food safety agency, has been determined at the level of

15%. It follows that both components—both the systematic component of 3% and random component of 4% of the recovery testing—can be included in the uncertainty budget. The complex uncertainty of the recovery is $u = \sqrt{32 + 42} = 5\%$.

Case 2: the recovery component is included in the uncertainty budget; the systematic component is used to introduce the recovery factor.

The content of the pesticide p, p′-DDE in the sausage has been determined. On the basis of the test of enriched samples (after they were minced), the recovery value was determined at the level of 60% + 8%. The value of the measurement uncertainty, stemming from the requirements of the food safety agency, has been determined at the level of 20%. It follows that including the systematic component (with a result 40% lower than the expected value), would cause the exceeding of the required value of uncertainty of 20%. With regards to that, the laboratory has used the recovery factor—that is, all the measurement results for real samples have been multiplied by the 1.6 factor and the variability of the recovery value of 8% was included in the uncertainty budget.

9.10.4 Range Control Chart

Another type of control chart is the range chart, on which the difference values between results for the two portions of the same control sample (most often analyzed in parallel) are recorded.

NOTE! If the program of internal quality management only provides for the evaluation of the results of parallel measurements of two sub-samples, then there is a danger that systematic errors will not be noticed.

The range chart is a one-sided card, which means that on the chart there is a determined line of the mean value for the calculated difference of two results and two upper lines of warning and action. The range chart is especially useful in the case of conducting tests for laboratory samples with a variable content of the analyte or with a changeable matrix content, and for non-durable samples in which the analyte concentration is rapidly changing.

9.11 How to Evaluate Control Charts

Interpretation of control charts is based on the assumption that the estimators of specific subsets (i.e. series of samples)—for example, the mean value—are changing randomly, rarely exceeding the control limits. The chart is divided into six areas. each 1 s wide, where s is the estimated standard deviation value of the test collective. The areas are placed symmetrically around the central line.

In cases where all points are between the internal control lines, the stability of the measuring system is assumed. However, if the points are more often on one side of

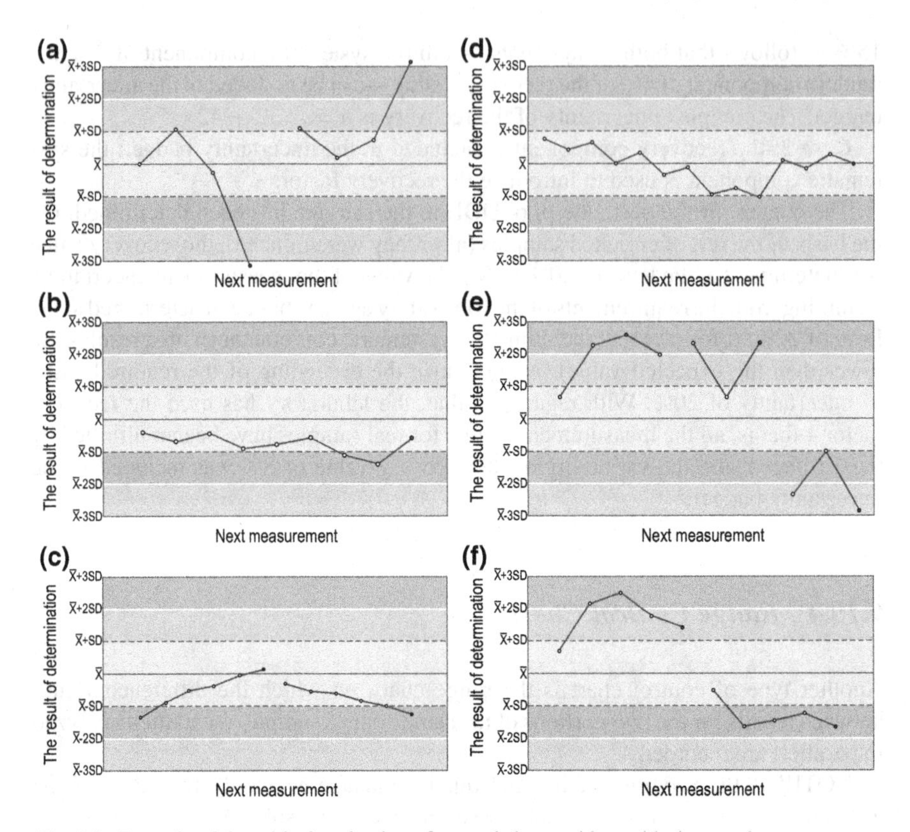

Fig. 9.2 Example of the critical evaluation of control charts with troubleshot results

the central line, it is suspected that a systematic error occurred, which could mean incorrect drawing of the central and control lines (Fig. 9.2).

Below the recommendation regarding the trend evaluation for results exceeding the specified criteria are shown (Fig. 9.2).

A. One result beyond the action line (additional measurements should be made) (Fig. 9.2a)
B. Nine consecutive results are on the same side of the center line (a systematic error occurred) (Fig. 9.2b)
C. Six consecutive results constantly increasing or decreasing (trend occurs) (Fig. 9.2c)
D. 14 consecutive results on both sides of the central line, of which the first five and the last four were conducted by one analyst and the middle four by another analyst (influence of the person who takes the measurements) (Fig. 9.2d)
E. Two out of four or three consecutive results are between the warning line and the action line (Fig. 9.2e)

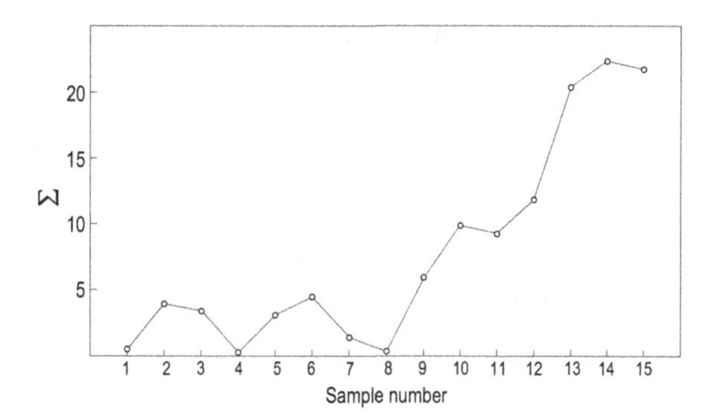

Fig. 9.3 Example of CumSum control chart

F. Four out of five consecutive results are between the warning line and the action line (Fig. 9.2f).

9.11.1 CuSum Control Chart

The name CuSum chart comes from the term cumulative sum, which means that during the chart creation consecutive sums are cumulated. CuSum chart is a control chart that uses the difference between the reference value and the measurement value ($X_{ref} - x$) for consecutive measurements so that the difference for a given measurement is added to the value determined for the previous measurements.

Other points are values corresponding to the sum of the differences for the selected number of measurements (Fig. 9.3).

The value that is the next point on the CuSum chart is calculated on the basis of the equation

$$C_i = \sum_{i=1}^{n} (\bar{x}_i - \mu) \quad (9.1)$$

where n is the number of consecutive results.

Figure 9.3 Example of CumSum control chart.

9.12 Summary

The objective of the statistical management of a process is to lead it to a stable, acceptable level (with acceptable variability) and keeping it on this level (within the acceptable limit of variability).

By using a CRM, information about the load of the measuring procedure can be obtained. By using the blank solution, information about the potential contamination of the measuring setup and/or reagents can be obtained. In the case of blank testing, it is recommended to introduce the value of the analytical signal and not the concentration value calculated from the calibration curve. In that way, the use of negative values, as well as values registered below the detectability limit, is justified. In cases where the blank does not hold the determined component, the central line should be at the zero level. Different placement of the central line (above zero) means that the blank was contaminated with the determined substance.

When planning the method of quality management in a laboratory, the fact that the CRMs might be an indispensable element of the management should be taken into account; however, they are not the only choice. An efficient quality management system should include regular use of materials of lower order, including: RMs, various kinds of control materials as well as homogeneous and well-characterized samples analyzed in the given laboratory. We also cannot forget about the programs that test the proficiency of the laboratory. Including all those elements in the systems leads to limiting the use of CRMs, and therefore to cost reduction.

Chapter 10
Interlaboratory Comparisons

Participation in interlaboratory comparisons allows demonstration of the laboratory's competences in the scope of conducted tests, supports the process of validation of analytical procedures, enables the comparison of results obtained in different laboratories, and under specified conditions, allows measurement traceability to be ensured.

According to the requirement of accreditation, all testing and calibrating laboratories should implement adequate quality control procedures for monitoring the quality of the reporting results. Those can be performed internally by using reference materials or conduct replicate tests with the same or different analytical procedures. Other means may focus on the externally organized interlaboratory comparison (ILC) or proficiency testing (PT), so as to obtain independent evidence regarding the competence of a given laboratory within its scope.

Interlaboratory comparison (ILC): organization, performance and evaluation of measurements or tests on the same or similar items by two or more laboratories or inspection bodies in accordance with predetermined conditions.

Clause 3.4; ISO/IEC 17043

Proficiency testing (PT): the evaluation of participant performance against pre-established criteria by means of interlaboratory comparison.

Clause 3.7; ISO/IEC 17043

The technical competence of the laboratory can be demonstrated by obtaining successful results in appropriate PT, but is can also be useful for participating in ILC focusing on broader purposes, for example: (1) to compare the results of two or more laboratories (ILC); (2) to evaluate the performance of a selected analytical procedure; and (3) to characterize a reference material.

The results of interlaboratory comparisons are always a valuable source of information for participating laboratories, since they allow an external evaluation of the

E. Bulska, *Metrology in Chemistry*, Lecture Notes in Chemistry 101,
https://doi.org/10.1007/978-3-319-99206-8_10

quality of the obtained results. Depending on the results obtained, participation in the ILC/PT can confirm high competencies of the laboratory in the scope of the specific testing. In cases where the results show any doubt, it can be a tool supporting the critical analysis of the potential problems. In principle, the participation of a laboratory in ILC can be voluntary, but there are some situations when it can be enforced by specific requirements (legal, accreditation, request from customer). In all cases where the evaluation of result is unsatisfactory, the laboratory should consider the cause of obtaining questionable or unsatisfactory results. This reason could arise from the systematic influence not being taken into consideration (e.g., incomplete recovery during extraction) or that that not all of the uncertainty components were considered during evaluation. Therefore, participation in interlaboratory comparisons should be considered as a meaningful tool allowing for the identification of problems. Thus, the critical evaluation of the results of ILC/PT supports all activities towards valid corrective actions.

In order to make the ILC/PT schemes useful for participating laboratories, the ILC/PT provider should use the well-defined and transparently described statistical procedure including the information on how the data will be processed and how the laboratory performance will be evaluated. This should also include information on how the assigned value/values for individual parameters will be assigned. The requirements for the ILC/PT providers are well described in ISO/IEC 17043, and the information on the statistical uses for the evaluation of laboratory results is discussed in ISO 13528.

A number of various approaches are known and are used to define the assigned value as well as the target range. This is essential, since the performance of individual laboratories is evaluated with scores, which are based on the comparison of laboratory results with an assigned value, dominated by the target range. In practice, various scoring systems are used, each with a specific evaluation criteria; thus is worthwhile to understand the idea behind the criteria.

The assigned value can be described in several ways, for example: based on formulation; as a certified reference value; and agreed as a consensus value.

Participation of laboratory in interlaboratory comparison.

Assessing the technical competencies of the laboratory, based on the result of the participation in ILC/PT, it should be evaluated whether the type of samples and applied measuring procedure correspond to the scope of measurements carried out in the laboratory. It rarely happens that a laboratory conducts measurements for one type of object only. Most often, the scope is diverse and encompasses various objects; their different properties are tested and various analytical techniques are used. Hence, it is important for the laboratory to skillfully gather those measuring procedures that can be evaluated with one type of interlaboratory comparison/proficiency testing.

The detailed information regarding the specific requirements is given in the EA-4/18:2010 '*Guidance on the Level and Frequency of Participation in Proficiency Testing.*' In the document, the choice of subdisciplines is also emphasized.

EA-4/18:2010
Level of participation: the number of sub-disciplines that an organization identifies within its scope, and therefore the number of specific proficiency tests that should be considered for participation.
Frequency of participation: how often a laboratory determines that it needs to participate in PT for a given sub-discipline; this may vary from sub-discipline to sub-discipline within a laboratory and between laboratories with the same sub-disciplines.
Sub-discipline: an area of technical competence defined by a minimum of one measurement technique, property and product, which are related.

Evaluation of the results of interlaboratory comparison/proficiency testing

Test and calibrating laboratories should strive to participate as often as possible in various ILC/PT, most of all because it is an incredibly important tool enabling the external evaluation of the quality of reporting results. Moreover, it is also possible to evaluate the reliability of the measuring procedures used. In each ILC/PT scheme, the main objective of the laboratory is to perform measurements for the object (sample) delivered by the provider, preferably under conditions typical to everyday practice in a given laboratory. Results obtained in the laboratory are then compared to the reference value assigned by the provider, and the evaluation of the laboratory performance includes the use of a selected score that classifies the given result in relation to the reference value.

Satisfactory result within the applied score: laboratory confirms its competencies in the scope of the measurements conducted.
Questionable/unsatisfactory result within the applied score: corrective action should be applied (the laboratory should conduct a detailed analysis of the measuring procedure used and the environmental conditions).

The main objective of participating in ILC/PT is to evaluate the performance of a given laboratory by comparing the result obtained by a laboratory (X_{lab}) with the reference value (X_{ref}) versus the accepted target range. As previously mentioned, the evaluation criteria should be provided to the participants by the ILC/PT provider. Thus, the laboratory can critically evaluate the statistical tools used by the organizer to reject the extreme values, to determine the reference values and to specify the evaluation criteria. All this information should be placed in the report, together with the results of the participating laboratories. That information is necessary for the critical assessment of the results obtained and, if the need arises, for using criteria stemming from legislative requirements or agreed upon with a customer. It is extremely important in view of the need to analyze the results of the comparisons (ILC/PT) by the laboratory, especially those that are doubtful and unsatisfactory. This explains the question as to what results should be qualified as doubtful or unsatisfactory and whether the criteria used by the organizers always meets the requirements regarding quality that are valid for the laboratory?

10.1 Assigning the Reference Value

The reference value, which is attributed to the quantity being measured in the samples distributed within the given ILC/PT round, can be determined in various ways. In practice, five scenarios are used by the organizers of ILC/PT.

10.2 By Formulation of Synthetic Test Samples

In this case, samples are used that are synthetically prepared, which means that the chemical substances of high purity are dosed by weighing the exact portion on the analytical balance. The known quantity of a pure substance can be mixed with other components by weight, which allows production of samples with a known content of the substance to be measured. A known quantity of a pure substance can be added to a natural sample (matrix enrichment), which allows production of samples mimicking the characteristics of test objects. The advantage of the production of synthetic samples by weight is the known quantity of the added substance with a great precision. It allows the measurement traceability to be determined and to assign an uncertainty to the reference value. Despite these advantages, in practice, it is not always possible to produce a synthetic material with expected metrological characteristics. Often the problem is the mixing of various components so as to obtain a satisfactory homogeneity of the sample. Moreover, it is not always possible to find a natural sample that does not contain the analyte. In addition, attention should be paid to the chemical form of a given substance as well as its affinity to the matrix.

10.3 The Use of Certified Reference Material

Although there is a limited number of available CRMs, whenever their matrix and the content of analyte correspond to that of test samples analyzed in the laboratory, they are considered to be the best option. The certified value for a given quantity, along with its uncertainty, is used as a reference (assigned) value. In that case, besides checking its competencies, the laboratory can also use the given CRM to ensure measurement traceability. From the metrological perspective, the use of a CRM is an ideal way to test the laboratory's performance. However, the high price of the CRM and their limited diversity means that they are seldom used by ILC/PT providers.

10.4 The Use of Reference Method

In cases where the reference method is used, the reference value is determined by either by a primary method or by a fully validated procedure, including calibration against a standard, with given traceability to national or international standard. An assigned value can be thus delivered by a single laboratory, analyzing the ILC/PT test item, and the target range is derived from the uncertainty of the performed measurement. This scenario is only possible when suitable CRM or chemical standards are available.

10.5 Consensus Value (Expert Laboratories)

In cases where the consensus value from expert laboratories is utilized, the results of measurements conducted by a group of expert laboratories are used. The individual laboratory is recognized as an expert laboratory, if it can prove its performance in the scope of a given determination (in a given sub-discipline) and if it uses validated measurement methods. Expert laboratories can be engaged in measurement conduct before the samples are sent to the participants of the proficiency testing or they can conduct tests in parallel, on-line with a given round of the PT.

The ILC/PT provider must account for the fact that not all results submitted by the expert laboratories are reliable, they can also be exposed to an unidentified systematic error. In that case, engaging a larger number of expert laboratories can limit the risk, as it is assumed to be highly unlikely that all laboratories would make such a mistake. It is, however, not always possible to find a larger number of expert laboratories with high technical competencies in a specific sub-discipline.

10.6 Consensus Value (from Participants)

In cases where the consensus value from participants is utilized, the assigned value is delivered as a consensus from the results of all participants, using the statistical methods described in ISO 13528, with consideration of the effects of outliers. The reference value is calculated as the mean value of the results submitted by the participants, after the rejection of deviating results or with the use of robust statistics. The most recommended are the Dixon tests (Q test) or Grubbs tests or the use of robust statistics in which the influence of the extreme values on the mean value is limited. Determination of the reference value on the basis of the results of all participants of the given round is a relatively simple and inexpensive method, hence it is readily used by the providers of PT. The method can be used for any type of test items, which allows for best fit of the sample type to the given sub-discipline. However, it should be remembered that the results obtained by the participants should be consistent enough for the reference value to be determined. Apart from that, there is always a risk of the presence of an unidentified systematic error. Additionally, such a method of operation does not ensure measurement traceability.

10.7 Evaluation of Laboratory Performance (Scores)

The main aim of all ILC/PT schemes is to provide external, objective evidence of the laboratory performance. For this purpose, the evaluation of the results given by a single laboratory is conducted by using various algorithms, by calculating the difference between the laboratory (X_{lab}) and the reference value (X_{ref}) versus the agreed denominator. These are named 'scores' and are specifically designed depending on the purpose. A number of different scoring algorithms are used by different ILC/PT providers; therefore, it is always recommended to be aware of the system used, before participation in given scheme.

In practice, the mathematical formula of a given score has a common numerator—the difference between the result obtained in a given laboratory (X_{lab}) and the reference value (X_{ref}) provided by the organizers.

$$\Delta = X_{\text{lab}} - X_{\text{ref}} \qquad (10.1)$$

where is X_{lab} is the result provided by the laboratory; X_{ref} is the reference value assigned by the provider of ILC/PT.

By using a different denominator, various scores can be defined. The most often applied and commonly accepted is the z-score; however, other evaluation criteria are also used.

10.8 Percentage Difference (*D*-Score)

The most straightforward way of comparing two numeric values is to determine the percentage of the difference, which in the case of ILC/PT can be given by following equation:

$$D_{\%} = X_{\text{lab}} - X_{\text{ref}}/X_{\text{ref}} \qquad (10.2)$$

where $D_{\%}$ is the difference expressed in percentage; X_{lab} is the result provided by the laboratory; X_{ref} is reference value.

The evaluation criteria are always determined by the ILC/PT provider; for example, it can be assumed that the satisfactory values meet the criteria $|D_{\%}| \leq 20\%$.

10.9 z-Score

In the case of using the ***z-score***, the following equation is used:

$$z = X_{\text{lab}} - X_{\text{ref}}/\sigma \quad (10.3)$$

The value (σ) in the denominator is the consensus value of the standard deviation for a given ILC/PT round. The reference value can be determined in various ways, described previously. It is also possible that (σ) can be determined as the spread of the results submitted by the participants, under the conditions of using robust statistics. This target range should be the accepted variability of results for the given measured quantity, most likely to fit for a given purpose.

NOTE: The target range value influences the performance score for an individual laboratory participating in ILC/PT.

Similar to the practice of setting the reference (assigned) value, several approaches can be used for setting the target range, used for evaluating the laboratory performance.

The prescribed value: the target range is set by the ILC/PT provider, so as to ensure that each laboratory is capable of reaching specific objectives of the PT; e.g. the meeting of legal requirements.

The expected value (by perception): the expected value (sometimes referred to as 'by perception') is a value determined by the organizer and the participating laboratories that reflects the expectations towards their performance.

The assigned value (by collaborative study): for the assigned value, the repeatability and reproducibility data from the interlaboratory validation of a given measuring procedure are used. In practice, this means that the laboratories are using the same measuring procedure and the same number of repetitions. The standard deviation of the results submitted by one laboratory reflects the within-laboratory precision (σ_R) and the standard deviation of results (mean values) submitted by the participating laboratories reflects the interlaboratory reproducibility (σ_L). Thus, the target range (σ) can be calculated as:

$$\sigma = \sqrt{(\sigma_L)^2 + (\sigma_R)^2/n} \quad (10.4)$$

where n is the number of replicate measurements.

10.9.1 The Assigned Value from Model Equation

In some situations, the value of the standard deviation used for the evaluation of the performance of a laboratory can be determined on the basis of a general model, for example, Horwitz function. So, it is possible to predict the standard deviation depending on the concentration of the determined analyte. The main limitation of this method is that the function does not always sufficiently reflect the real reproducibility of the measuring procedure achieved practically by the laboratory.

10.9.2 The Assigned Value from PT Round

In this case, the measurement results from all participants of a given round are used. The standard deviation of the mean value is calculated with the use of robust statistics, in which the influence of the extreme values on the mean value is limited.

When ***z-score*** is used, following evaluation criteria are used:

$|z| \leq 2.0$ satisfactory performance
$2.0 < |z| < 3.0$ questionable performance
$|z| \geq 3.0$ unsatisfactory performance.

Another way to measure the performance of the laboratory is to use the ***z′-score***, which is similar to ***z-score***, but with an extended denominator. In the equation describing the ***z-score***, the uncertainty of the reference value is not included. Thus, ***z′-score*** has been proposed—one that includes both the target range (as included in the equation for ***z-score***) and the standard uncertainty of the reference value (u_X).

$$z' = X_{\text{lab}} - X_{\text{ref}}/\sqrt{\sigma^2 + u_X^2} \quad (10.5)$$

where u_X stands for the standard uncertainty for the reference value.

The criteria for the evaluation of results with the use of the ***z′-score*** is the same as in the case of the ***z-score***.

When ***z′-score*** is used, the following evaluation criteria are used:

$|z'| \leq 2.0$ satisfactory performance
$2.0 < |z'| < 3.0$ questionable performance
$|z'| \geq 3.0$ unsatisfactory performance.

NOTE! The value of the *z′-score* will significantly differ from the value of *z-score* if the uncertainty value of the laboratory is significant.

10.10 Zeta Score

Since both ***z-score*** and ***z′-score*** do not consider the uncertainty of the results reported by single laboratory, another extension in denominator leads to ***zeta score***, which includes the standard uncertainty of the result received from the laboratory. The use of ***zeta score*** is only possible when the laboratory provides the result with an assigned uncertainty. It is worthwhile to emphasize that the ILC/PT providers often ask for results together with the uncertainty, which stems from the recommendations of the ISO/IEC 17043. The evaluation criteria are identical to the previously stated criteria for the z- and z'-scores.

$$\text{zeta} = \frac{x_{\text{lab}} - X_{\text{reference}}}{\sqrt{u_x^2 + u_X^2}} \quad (10.6)$$

NOTE! *zeta score* cannot be used when the reference value is obtained from the results of the ILC/PT participants.

10.11 E_n Score

A less frequently used criterion is E_n scores, known also as normalized error. This score is useful for evaluation the results of a single laboratory versus reference value.

$$E_n = X_{\text{lab}} - X_{\text{pt}} / \sqrt{U_{\text{lab}}^2 + U_{\text{pt}}^2} \quad (10.7)$$

where U_{lab} and U_{pt} are the values of the expanded uncertainty (for $k = 2$) for the result submitted by the laboratory and for the reference value, respectively.

It is worth noting that in the case of the En number, the expanded uncertainty values are inserted as denominator, hence this influence the evaluation criteria.

$|En| \leq 1.0$ satisfactory performance
$|En| > 1.0$ unsatisfactory performance.

10.12 Summary

The above discussion on the most important scores used by the ILC/PT providers indicates that the evaluation of the performance of a single laboratory depends on the applied criteria. Hence, a very important part of the reports is the inclusion of a clear description of the used scores. However, it is also important to highlight that the laboratory, when evaluating its participation in the given comparison, should critically evaluate the extent to which the criteria used correspond to the expected requirements regarding the scope of the activity.

Example

In the ILC aiming for determination of a selected substance in soil, there were 25 participants. Of these, 20 submitted their results with an assigned uncertainty. In this case, the results provided by the laboratory were compared to the certified reference values, since the test item was a CRM. NOTE: the results listed in the table are raw data as reported by laboratory, without rounding.

The certified content of the substance of interest is 0.51 mg/L $\pm$ 0.048 mg/L, for $k = 2$.
The standard uncertainty of the reference value is $u_{\mathrm{ref}} = 0.024$ mg/L.
The calculated standard deviation of the mean value from the results of all participants is $\sigma = 0.034$ mg/L.

Results submitted by the participants were statistically evaluated according to ISO/DIN 13528 standard '*Statistical Methods For Use in Proficiency Testing by Interlaboratory Comparisons.*' The results submitted by the laboratories and the scores for each laboratory are shown in Table 10.1.

Column I: ID of the laboratories participating in the comparison.
Column II and III: the results submitted by the laboratories (value and uncertainty, respectively).
Column IV–VII: selected scores.

NOTE! Values of the ***zeta score*** and ***E_n score*** were calculated only for those laboratories that submitted their results with an assigned uncertainty.

10.12.1 Pay Attention to the Results of the Laboratory with ID 22

When z and z'-scores were used, in both cases their values indicated unsatisfactory performance of the laboratory 22.

When zeta score and E_n score were used, in both cases their values indicated satisfactory performance of the laboratory 22 (zeta < 2; $E_n \leq 1$).

With this example, taken from real practice, considering the results reported by participating laboratories, it can be seen how the selection of adequate evaluation criteria (scores) is essential for the evaluation of a given laboratory. The example of laboratory 22 shows that the evaluation of the result on the basis of indicators that only take into consideration the uncertainty assigned to the reference value is harsher

Table 10.1 Evaluation of the results from the laboratory participating in ILC round

I	II	III	IV	V	VI	VII
Lab ID	Mean value as delivered by lab	Standard uncertainty*, u	z score	z' score	zeta score	E_n score
1	0.53	0.015	0.59	0.50	0.71	0.35
2	0.494	–	−0.47	−0.40	–	–
3	0.503	0.086	−0.21	−0.18	−0.08	−0.04
4	0.51	0.0255	0.00	0.00	0.00	0.00
5	0.482	0.011	−0.82	−0.70	−1.06	−0.53
6	0.5	0.06	−0.29	−0.25	−0.15	−0.08
7	0.557	0.0557	1.38	1.18	0.77	0.39
8	0.579	–	2.03	1.73	–	–
9	0.484	0.028	−0.76	−0.65	−0.71	−0.35
10	0.495	0.014	−0.44	−0.38	−0.54	−0.27
11	0.515	0.027	0.15	0.13	0.14	0.07
12	0.5	0.075	−0.29	−0.25	−0.13	−0.06
13	0.534	0.002	0.71	0.60	1.00	0.50
14	0.586	–	2.24	1.90	–	–
15	0.5	0.05	−0.29	−0.25	−0.18	−0.09
16	0.519	0.088	0.26	0.23	0.10	0.05
17	0.519	0.0519	0.26	0.23	0.16	0.08
18	0.502	–	−0.24	−0.20	–	–
19	0.483	0.058	−0.79	−0.68	−0.43	−0.22
20	0.49	–	−0.59	−0.50	–	–
21	0.546	0.00819	1.06	0.90	1.42	0.71
22	**0.599**	**0.038**	**2.62**	**2.23**	**1.98**	**0.99**
23	0.509	0.009	−0.03	−0.03	−0.04	−0.02
24	0.529	0.04761	0.56	0.48	0.36	0.18
25	0.49	0.03	−0.59	−0.50	−0.52	−0.26

*For those laboratories that submitted the result with an expanded uncertainty for $k = 2$, the value was divided by 2

The critical example of the laboratory 22 is highlighted in bold

in comparison to the evaluation based on indicators that also include the uncertainty provided by the laboratory. Depending on the scope of the laboratory activities as well as the purpose of measurements, the use more demanding or fit for purpose scores for the evaluation of laboratory performance is justified.

When discussing the variety of evaluation criteria, it is reasonable to ask why the ILC/PT providers do not use one commonly accepted score. One point to consider is that, according to the requirements of the ISO/DIS 13528, the ILC/PT provider has a right to apply criteria (scores) depending on the needs of a given group of laboratories. This allows the laboratory performance to be evaluated with regard the expected use of the result.

In the case, when the reference value is determined with the primary method, for example ID ICP MS (Isotopic Dilution Inductively Coupled Plasma Mass Spectrometry), the uncertainty assigned to the reference value is very small. In laboratories working in the area of the environmental analysis, a higher uncertainty is most often satisfactory, hence it was decided that taking into consideration the uncertainty provided by the laboratory would ensure a more reliable method of evaluation.

10.12.2 Indicators of the Evaluation of Laboratories Participating in Interlaboratory Comparisons

$$z = \frac{(x - X_{\text{odn}})}{\sigma_p}$$

$$z' = \frac{(x - X_{\text{odn}})}{\sqrt{\left(\sigma_p^2 + u_X^2\right)}}$$

$$zeta = \frac{x - X_{\text{odn}}}{\sqrt{u_x^2 + u_X^2}}$$

$$E_n = \frac{x - X_{\text{odn}}}{\sqrt{U_x^2 + U_X^2}}$$

10.13 Summary

The evaluation criteria used in ILC/PT allow the evaluation of the performance of a single laboratory, whether or not it meets the preset criteria in respect of the reported results. The evaluation scores z and z' do not include the uncertainty of the result reported by the given laboratory, whereas zeta and E_n scores accommodate it in the denominator of respective equations.

Obtaining satisfactory performance with z and z'-scores (at the same time) means that the laboratory meets the requirements set by the organizer of ILC/PT; obtaining a satisfactory performance of the zeta and E_n scores means that the laboratory meets its own criteria. Various situations, when using z-score and zeta score are summarized in Table 10.2, showing the evaluation criteria in practice.

Table 10.2 Summary of the evaluation of the performance of laboratory

Performance scores		Laboratory self-evaluation		Comments
z-score	zeta score	ILC/PT provider evaluation	Laboratory self-evaluation	
Yes	Yes	Yes	Yes	Satisfactory performance
Yes	No	Yes	No	Underestimated uncertainty
No	Yes	No	Yes	Uncertainty of the results reported by laboratory exceed the uncertainty accepted by ILC/PT provider
No	No	No	No	Unsatisfactory performance

Index

E. Bulska, *Metrology in Chemistry*, Lecture Notes in Chemistry101,
https://doi.org/10.1007/978-3-319-99206-8

Printed in the United States
By Bookmasters